Springer Series on
ATOMIC, OPTICAL, AND PLASMA PHYSICS 34

Springer-Verlag Berlin Heidelberg GmbH

Physics and Astronomy ONLINE LIBRARY

http://www.springer.de/phys/

Springer Series on
ATOMIC, OPTICAL, AND PLASMA PHYSICS

The Springer Series on Atomic, Optical, and Plasma Physics covers in a comprehensive manner theory and experiment in the entire field of atoms and molecules and their interaction with electromagnetic radiation. Books in the series provide a rich source of new ideas and techniques with wide applications in fields such as chemistry, materials science, astrophysics, surface science, plasma technology, advanced optics, aeronomy, and engineering. Laser physics is a particular connecting theme that has provided much of the continuing impetus for new developments in the field. The purpose of the series is to cover the gap between standard undergraduate textbooks and the research literature with emphasis on the fundamental ideas, methods, techniques, and results in the field.

27 **Quantum Squeezing**
By P.D. Drumond and Z. Spicek

28 **Atom, Molecule, and Cluster Beams I**
Basic Theory, Production and Detection of Thermal Energy Beams
By H. Pauly

29 **Polarization, Alignment and Orientation in Atomic Collisions**
By N. Andersen and K. Bartschat

30 **Physics of Solid-State Laser Physics**
By R.C. Powell
(Published in the former Series on Atomic, Molecular, and Optical Physics)

31 **Plasma Kinetics in Atmospheric Gases**
By M. Capitelli, C.M. Ferreira, B.F. Gordiets, A.I. Osipov

32 **Atom, Molecule, and Cluster Beams II**
Cluster Beams, Fast and Slow Beams, Accessory Equipment and Applications
By H. Pauly

33 **Atom Optics**
By P. Meystre

34 **Laser Physics at Relativistic Intensities**
By A.V. Borovsky, A.L. Galkin, O.B. Shiryaev, T. Auguste

35 **Many-Particle Quantum Dynamics in Atomic and Molecular Fragmentation**
Editors: J. Ullrich and V.P. Shevelko

Series homepage – http://www.springer.de/phys/books/ssaop/

Vols. 1–26 of the former Springer Series on Atoms and Plasmas are listed at the end of the book

A.V. Borovsky A.L. Galkin
A.B. Shiryaev T. Auguste

Laser Physics at Relativistic Intensities

With 85 Figures

Springer

Professor Dr. Sci. Andrew V. Borovsky
Automation and Control Systems Laboratory
IrkutskEnergo Power Company
Sukhe-Bator Street 3
664000 Irkutsk, Russia
e-mail: avb@irk.ru

Dr. Sci. Andrew L. Galkin
Ph.D. Oleg B. Shiryaev
General Physics Intitute of the Russian Academy of Science
38 Vavilov St., Box 117942, Moscow, Russia
e-mail: galkin@kapella.gpi.ru
e-mail: obs@kapella.gpi.ru

Dr. Thierry Auguste
Commissariat à l'Energie Atomique
DSM/DRECAM/SPAM
Bât. 522 CE Saclay
91191 Gif-sur-Yvette Cédex, France
e-mail: auguste@drecam.cea.fr

DOI 10.1007/978-3-662-05242-6

Library of Congress Cataloging-in-Publication Data.

Laser physics at relativistic intensities / A. V. Borovsky... [et al.].
p. cm. - (Springer series on atomic, optical, and plasma physics, ISSN 1615-5653; 34)
Includes bibliographical references and index.

1. Laser-plasma interactions. 2. Laser pulses, Ultrashort. 3. High power lasers.
I. Borovskii, A.V. (Andrei Viktorovich) II. Series.
QC718.5.L3 L37 2002
530'.4'46-dc21 2002021783

http://www.springer.de

Originally published by Springer-Verlag Berlin Heidelberg New York in 2003
MyCopy version of the original edition 2003

Typesetting by the authors
Final page layout: EDV-Beratung Frank Herweg, Leutershausen
Cover concept by eStudio Calmar Steinen
Cover design: *design & production* GmbH, Heidelberg

Printed on acid-free paper SPIN: 10752510 57/3141/tr - 5 4 3 2 1 0
www.springer.com/mycopy

Preface

One of the major accomplishments of laser technology that took place during the last 15 years is the possibility of generating coherent radiation that can be focused so that its intensity reaches the magnitude of $10^{18}\,\mathrm{W/cm^2}$. Even higher intensities result from nonlinear self-focusing of such radiation in matter. The unique character of these magnitudes is illustrated by the fact that these intensities are substantially higher than those occurring inside the Sun. A range of previously unexplored physical mechanisms come into play as laser pulses interact with matter under the conditions of an extreme concentration of laser energy. In particular, free electrons of plasma formed by rapid nonlinear ionization of gases and solid targets during the pulse rise time are driven by a laser radiation electric field at velocities comparable to the speed of light, and the corresponding relativistic increase in their masses entails a modification of the plasma's optical properties. Laser radiation intensities at which the above effect occurs are called relativistic. This book is intended to provide an introduction to the field of laser physics at relativistic intensities.

Extensive theoretical and experimental studies have been performed in this area in the last decade. At present, laser physics at relativistic intensities can be considered a new and rapidly evolving area of modern physics. Important basic new phenomena and concepts of applications are associated with it, among them relativistic and charge-displacement self-channeling, wakefield particle acceleration, generation of free electron harmonics and X-rays, and fast ignition.

In the introductory chapter of this book, we discuss the basic physical mechanisms determining the character of propagation of superintense laser radiation in plasmas.

The theoretical part of the book begins in Chap. 2 with a brief introduction to cold plasma electrodynamics that serves as the basic theoretical model of interactions of ultrashort, high-intensity laser pulses with matter.

The complexity of theoretical studies of laser–plasma interactions at relativistic intensities is due to the fact that they can be described adequately only in the framework of fully nonlinear approaches . In Chap. 3, we develop a class of traveling wave solutions of electron fluid dynamics and Maxwell equations (solutions of the Akhiezer–Polovin problem) corresponding to plane electromagnetic waves of arbitrary intensities in plasmas and their nonlinear amplitude and phase self-modulation.

The instabilities of the above solutions are treated in Chaps. 4 and 5 with the help of a special formalism developed to calculate growth rates for linearized equations with oscillating coefficients. This theory describes the scattering of superintense electromagnetic radiation by plasmons, the fluid dynamics analog of the Compton effect, the excitation of harmonics, and the emergence of a continuum of scattered radiation. Note that laser radiation scattering can serve as a mechanism for experimentally diagnosing the propagation of relativistically intense laser pulses in plasmas.

Fluid dynamics of the plasma electron component and Maxwell equations are also used in Chap. 6 to establish a hierarchy of conservative models of propagation of superintense, ultrashort laser pulses in cold underdense plasmas. These models describe laser radiation diffraction, refraction, and dispersion; relativistic and charge-displacement nonlinearities, and the excitation of plasmons by propagating pulses. Large aperture and long laser beam limits of the above models are considered. Each of the above models is of certain interest from the point of view of the theory of nonlinear waves.

Analytical and numerical studies of the propagation of superintense, ultrashort, large aperture laser pulse solitons in plasmas, some of which have been performed in the framework of the models developed, are summarized in Chap. 7.

Chapters 8 and 9 are dedicated to a new basic regime of propagation of superintense laser beams in cold underdense plasmas, called relativistic and charge-displacement self-channeling. A theory and results of simulations of this phenomenon using models based on nonlinear Schroedinger and wave equations are presented. The intensities associated with channeled propagation are extremely high and can reach the level of 10^{21} W/cm^2, and this fact is the basis of a method of concentrating laser radiation energy in matter. The stability of the relativistic and charge-displacement self-channeling regimes is also examined.

Results on the multiphoton ionization of matter and the propagation of light in gases governed by the nonlinearity stemming from the above effect are presented in Chap. 10.

Experimental studies of laser–matter interactions at high intensities including relativistic and charge-displacement self-channeling, the enhancement of the channeled propagation distance by an exterior energy supply, X-ray laser schemes, experiments with harmonics excitation, the acceleration of electrons, and generation of superintense magnetic fields are described in Chap. 11.

This book is addressed to scientists working in the areas of high-intensity laser physics, nonlinear plasma waves, and nonlinear optics as well as physics and electrical engineering graduate students. We assume that our reader is familiar with the basic concepts of plasma physics.

This book is largely based on original results obtained by its authors. Consequently, to some extent, the choice of topics in it reflects our own research interests. But in fact, it is not possible for any monograph to be comprehensive when the subject is such an area of vigorous research that new results emerge continuously and conflicting opinions are frequent. A bibliography is presented to provide additional sources of information for readers interested in a broader range of aspects of laser–plasma interactions at relativistic intensities or alternative approaches to the problems discussed.

We are delighted that here we have an opportunity to acknowledge the cooperation of all of the scholars with whom we joined efforts working on the problems discussed below. We express our special gratitude to Prof. V.V. Korobkin (General Physics Institute of the Russian Academy of Science), Dr. A.B. Borisov (University of Illinois at Chicago), Prof. C.K. Rhodes (University of Illinois at Chicago), Prof. R.N. Sudan (Cornell University), and Dr. P.V. Nickles (Max Born Institute for Nonlinear Optics and Short-Time Spectroscopy).

Moscow, Gif-sur-Ivette
May 2003

Andrew Borovsky, Andrew Galkin
Oleg Shiryaev, Thierry Auguste

Contents

1. Introduction

1.1 The Subject of Laser Physics at Relativistic Intensities

Laser physics originates from the fundamental work performed by N.G. Basov, A.M. Prokhorov, and C.H. Townes in the late 1950s. At present, it is an extensive and rapidly evolving area of knowledge of the nature of coherent electromagnetic fields; the electronic structure of matter; and interactions of electromagnetic fields with gases, solids, liquids, and plasmas. Modern laser physics also comprises such engineering disciplines as laser design and technology, means of laser radiation control, and the development of various experimental hardware based on the use of laser light. Laser radiation is characterized by four basic parameters: wavelength λ, pulse duration τ, intensity I, and power P. This part of laser physics which can be termed "traditional" has been studied for over 30 years. It deals with the following ranges of the above parameters: $\lambda > 100\,\mathrm{nm}$, $\tau > 1\,\mathrm{ps}$, $I < 10^{16}\,\mathrm{W/cm^2}$, $P < 10^9\,\mathrm{W}$. A number of monographs have been dedicated to this area of physics [1–3]. The past decade has witnessed notable "knowledge saturation" in this field, and to a large extent traditional laser physics is gradually becoming a subject of applied rather than basic studies as its achievements materialize in specific hardware for various technologically advanced human activities.

At the same time, during the last 10 years this process was paralleled by rapid progress in basic research in a novel area of laser physics, where the ranges of laser radiation parameters are $1 < \lambda < 100\,\mathrm{nm}$, $10^{-3} < \tau < 1\,\mathrm{ps}$, $10^{16} < I < 10^{22}\,\mathrm{W/cm^2}$, and $10^9 < P < 10^{13}\,\mathrm{W}$. Laser pulses characterized by such parameters are generated with the help of the powerful excimer, Ti:sapphire, and Nd-glass laser systems developed in a number of leading laser physics laboratories throughout the world. These lasers are designed according to the chirp pulse amplification (CPA) technique [4, 5]. Figure 1.1 summarizes the main laser systems with power beyond 1 TW.

Rapid nonlinear ionization occurs early during the pulse rise time when radiation with an intensity of such magnitude is focused on matter (the ionization threshold is approximately $10^{14}\,\mathrm{W/cm^2}$), and the central part of the pulse propagates into the resulting plasma. Electromagnetic field intensities at which the velocities of plasma free electrons driven by the electromagnetic

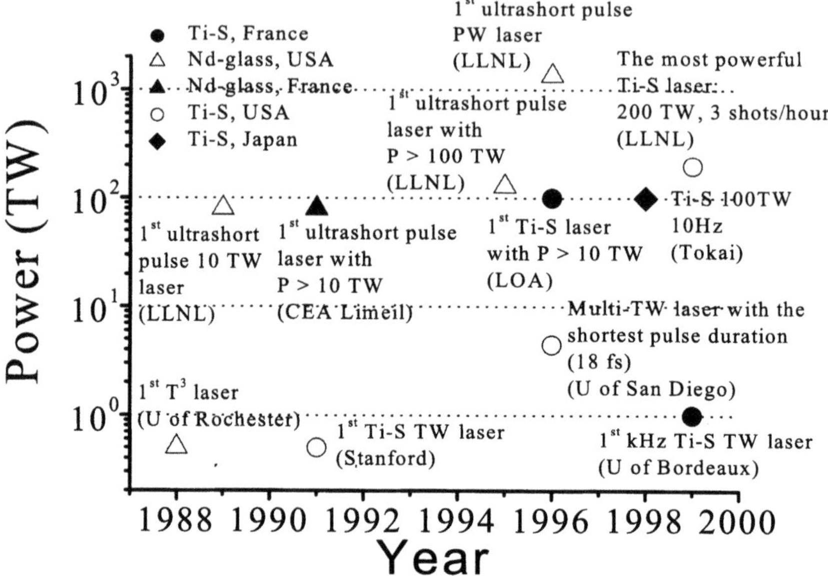

Fig. 1.1. Evolution of laser power during the last 10 years

field become comparable to the speed of light are called *relativistic*. This book is dedicated to several major issues in laser physics at relativistic intensities.

To evaluate relativistic intensity, consider the relativistic mass factor $\gamma = \sqrt{1+(\boldsymbol{p}/mc)^2}$ for an electron driven by an electromagnetic wave. Here $\boldsymbol{p}$ and m are the electron momentum and rest mass, and c is the speed of light. From canonical momentum conservation, $\boldsymbol{p} = (e/c)\,\boldsymbol{A}$, where $\boldsymbol{A}$ is the electromagnetic field vector potential. For a circularly polarized wave given by

$$\boldsymbol{A} = \frac{\boldsymbol{e}_1 + \mathrm{i}\boldsymbol{e}_2}{2}\, a_0 \exp[\mathrm{i}(kx_3 - \omega t)] + c.c.\,,$$

the γ factor is expressed as a function of the pump intensity $\gamma = \sqrt{1 + I/I_\mathrm{r}}$, where the parameter I_r is

$$I_\mathrm{r} = \frac{m^2\omega^2c^3}{4\pi e^2} = 2.75 \times 10^{18} \left(\frac{1}{\lambda[\mu\mathrm{m}]}\right)^2 (\mathrm{W/cm^2})\,. \tag{1.1}$$

Obviously, relativistic effects in the motions of laser driven plasma free electrons are significant when the electromagnetic radiation intensity is comparable to I_r. For this reason, in a narrower sense, this wavelength-dependent parameter is defined as the *relativistic intensity*.

Generally physics of interaction of superintense electromagnetic radiation with matter has three interrelated aspects:

1. **Physics of elementary atomic processes in superintense optical fields.** The pertinent phenomena include
   - nonlinear field-induced ionization of matter during the pulse rise time, resulting in plasma formation;
   - polarization of plasma ions by exterior fields (Kerr effect);
2. **Optics of laser pulse propagation in matter.** The basic phenomena of this type are
   - laser radiation diffraction;
   - laser radiation refraction in a medium with a variable refractive index distribution;
   - laser radiation dispersion.
3. **Nonlinear physics of plasmas in intense electromagnetic fields.** Plasma optical properties are modified by the following mechanisms:
   - *Relativistic nonlinearity.* This effect plays a central role in the phenomena considered in this book. The local value of the plasma frequency and, therefore, the value of its refractive index depend on the mass of plasma free electrons. As discussed above, electromagnetic radiation of extremely high intensity drives them at velocities comparable to the speed of light, causing an increase in their masses and, consequently, resulting in a local intensity-dependent variation of the plasma's refractive index.
   - *Charge-displacement nonlinearity.* This phenomenon is due to the perturbation of the plasma electron concentration by the ponderomotive force, which also entails a local plasma refractive index variation.
   - *Excitation of waves in the plasma electron component by the propagating radiation.* Although the intense electromagnetic field displaces plasma electrons, the inertially frozen positive plasma ion background provides a restoring force, thus causing electron fluid oscillations. An important related effect is Raman scattering of intense electromagnetic radiation by plasmons that it excites during propagation.
   - Laser pulse energy dissipation due to ionization, collisional and multiphoton absorption, etc.
   - Inertial nonlinearities such as ion motions, heat transfer, etc.
   - Quasi-static magnetic field generation by electron motions.

The above brief outline of the range of pertinent physical mechanisms illustrates the complexity of the problem of laser–matter interactions at relativistic intensities. Naturally, under various conditions, substantial model simplifications are possible, as will be discussed in the following chapters. First of all, note that since ionization occurs rapidly on the laser pulse front, the studies presented below deal with the propagation of the "main" temporal part of the pulse in a plasma which is considered preformed. The propagation nonlinearity owing to multiphoton ionization is considered separately in

Chap. 10 of this book. Furthermore, in the case of interaction of relativistically intense laser radiation with light gases, molecules are ionized completely, which eliminates the need for considering the Kerr effect. Therefore, in this book, we focus on effects related to nonlinear optics and plasma physics. Readers can find a review of intense laser radiation interaction studies under the conditions for which the dominant role is played by atomic processes in [6].

Besides, inertial nonlinearities do not play a significant role in the propagation of ultrashort pulses whose durations are much smaller than the plasma response time and can also be excluded from consideration.

Studies of relativistic intensity laser–matter interaction have already revealed a whole range of novel and physically meaningful phenomena. Probably the most remarkable of them is relativistic and charge-displacement self-channeling which was predicted theoretically in [7] and later observed experimentally [8]. It was demonstrated that under certain conditions, relativistically intense laser beams tightly focused into gases propagate in spatially confined modes and produce channels spanning over 100 Rayleigh lengths. Since the laser beam intensity is maximal near the propagation axis, the combined effect of the relativistic electron mass increase and ponderomotive electron component expulsion tends to reduce the plasma frequency in the paraxial domain and thus enhance medium transparency in it compared with the beam periphery, which results in diffraction suppression and propagation in the self-channeling regime. Furthermore, cavitation occurs after a sharp intensity increase at the propagation axis due to beam nonlinear self-focusing; all of the electrons are pushed out of the paraxial domain by the ponderomotive force, and the propagating laser radiation is trapped in the resulting channel. A unique self-concentration of laser radiation energy in matter is associated with relativistic and charge-displacement self-channeling: electromagnetic fields become comparable with thermonuclear values, and corresponding intensities can be as high as $10^{22}\,\mathrm{W/cm^2}$. Relativistic and charge-displacement self-channeling regimes are possible when the laser pulse power is greater than the critical power of relativistic self-focusing which is expressed as [7]

$$P_{\mathrm{cr}} = 16.2\,(\omega/\omega_{\mathrm{p}})^2\,\mathrm{GW}\,. \tag{1.2}$$

Here, $\omega_{\mathrm{p}} = \sqrt{4\pi e^2 n/m}$ is the plasma electron frequency, ω is the laser pulse frequency, and n is the unperturbed plasma electron concentration.

In accord with the authors' research interests, a semianalytical approach is adopted in the theoretical part of this book. Fundamental plasma electrodynamics equations are used below as a basis for establishing simplified models of laser–plasma interactions at relativistic intensities, and these interactions are treated in the model framework. Special attention is also paid to partial solutions that can be obtained analytically or with a minimal amount of computations and to their stability theory. In our opinion, the advantages of this method are not limited to the possibility of avoiding extremely large computations. The emphasis on analytical techniques makes it possible to identify clearly the most important specific physical mechanisms relevant to every particular powerful laser–plasma interaction phenomenon studied. Besides, we hope that due to this approach, this book will be more useful to graduate students as an introduction to the theory of laser plasma interactions at relativistic intensities since it is possible to follow all the derivations given below. An alternative approach, which is used extensively in the studies of laser–plasma interactions, is to generate new knowledge with the help of massive particle-in-cell (PIC) simulations. Some of the major results of such studies are also reflected in this book.

1.2 A Review of Basic Studies of Laser Physics at Relativistic Intensities

At present, an extensive literature exists on laser–matter interactions at high intensities and various related issues. Some aspects of this problem are reflected in monographs [9–11]. A systematic presentation of relativistic intensity laser physics, including a theory of relativistic and charge-displacement self-channeling can be found in [12]. Studies relevant to the theory of interactions of relativistically intense laser pulses with plasmas can be divided into the following groups.

1.2.1 Nonlinear Propagation and Self-Focusing of Light in Matter

Naturally, many of the basic concepts and models of relativistic intensity laser physics stem from the ideas of "conventional" nonlinear optics. For this reason, some of the classical laser radiation propagation studies are relevant to the subject of this book.

Pioneering research into nonlinear self-focusing of light was carried out by G.A. Askar'yan. The very possibility of this phenomenon was predicted in [13]. This paper also treated thermal and ponderomotive charge-displacement nonlinearities in laser-irradiated plasmas. The first experimental observations of self-focusing were reported in [14, 15].

A nonlinear Schroedinger equation which serves as the basic model of optical field evolution in nonlinear media was derived in [16, 17].

Nonlinear media, where the refractive index variation is proportional to the intensity of the propagating optical field, are called cubic. Slab geometry solitons and axially symmetrical lowest eigenmodes for the corresponding Schroedinger equation (also called cubic) were developed in [18]. A countable set of cubic Schroedinger equation axially symmetrical eigenmodes was presented in [19, 20].

Multisoliton solutions to the slab geometry cubic Schroedinger equation were found in [21] using the inverse scattering method. The asymptotics of this equation for large times were developed in [22].

Plane wave instability and filamentation of light in cubic media were investigated in [23].

Nonlinear Schroedinger equation eigenmode stability was examined in a large number of papers [24–26]. A special variational technique was proposed for this purpose [24] and applied to prove the linear stability of the lowest eigenmode for the class of nonlinear Schroedinger equations with saturating nonlinearities [26].

It was also shown that an exact self-similar solution of the cubic Schroedinger equation exists [27]. An important effect associated with the cubic Schroedinger equation of a sufficiently high dimension is the Schroedinger blowup: when the system Hamiltonian is positive, solutions become infinite in a finite time interval. The proof of the above statement for a generalized Ginzburg–Landau type equation can be found in [28]. Numerous studies were dedicated to the Schroedinger blowup. A hypothesis according to which the Schroedinger blowup has a self-similar character was proposed and verified numerically in [29, 30], where blowup rates calculated on the basis of the self-similar solution of the cubic Schroedinger equation were also presented.

Article [31] constituted an important step toward understanding the possibility of channeled propagation in media with saturating nonlinearities. As shown in this work, the Hamiltonian of the corresponding Schroedinger equation must be negative to provide for optical field nonlinear self-trapping (optical guiding), whereas diffraction prevails and beam spreading occurs when the Hamiltonian is positive.

1.2.2 Charged Particle Motions in Electromagnetic Fields

The well-known solution of the problem of the motion of a charged particle in a monochromatic field can be found in [32]. Analytic solutions for the long pulse limit were established in [33]. In another study [34], obviously performed in anticipation of the development of powerful sources of coherent electromagnetic fields, a classical theory was developed describing Thomson scattering by free electrons in idealized free space plane waves. This theory accounted for harmonics production and included radiation spectra and angular distribution expansions in pump wave intensity. Later, these issues were reexamined in [35]. Radiated energy and radiation reaction for a relativistic electron in a plane electromagnetic wave of an arbitrary amplitude were also discussed in [36]. Additional analysis and numerical solutions describing charged particle motions in high intensity optical fields can be found in [37, 38]. Experimental evidence of second- and third-harmonics generation in the relativistic regime of interaction was recently obtained [298].

1.2.3 Nonlinear Electromagnetic Waves in Plasmas

The classic paper of A.I. Akhiezer and R.V. Polovin [39] published in 1956 is an outstanding study of nonlinear electromagnetic waves in plasmas. A fully nonlinear set of equations describing traveling waves was derived from slab geometry relativistic fluid dynamics and Maxwell equations. This theory accounts for the relativistic nonlinearity of electromagnetic wave propagation and for the coupling of electromagnetic waves and plasmons. Important exact and approximate solutions of the above set of equations were also developed. It is remarkable that this paper, so important for understanding laser–plasma interactions at relativistic intensities, was published prior to the invention of lasers.

Several studies were dedicated to the set of traveling wave equations derived by A.I. Akhiezer and R.V. Polovin [40–42]. In [43] it was used to study, analytically and numerically, the propagation of electromagnetic wave–plasmon formations in plasmas with phase velocities close to the speed of light and their amplitude and phase self-modulation (also, see [44]).

A problem somewhat similar from the mathematical point of view was studied in [45], where slab geometry relativistically intense solitons moving in cold underdense plasmas with quantized group velocities were found numerically. Later the same results were obtained by other authors [46]. An analytical approach to the relativistically intense soliton problem was proposed in [47].

Papers [48, 49] were dedicated to charged-particle acceleration by beat wave accelerators.

Various laser plasma phenomena are dealt with in [50]. In particular, a definition of relativistic intensity is introduced in this monograph (naturally, this value is present implicitly in [39]). This definition is based on the assumption that at relativistic intensity, the energy of the electron's field-induced oscillations equals $m_{e,0}c^2$ and as a result the relativistic intensity definition of [50] differs from that adopted in this book by a factor of 3.

Paper [51] is of special interest among the studies of relativistic plasma fluid dynamics. It treats a number of plasma models in canonical variables in the framework of Hamiltonian formalism, including the quasi-linear perturbation theory.

1.2.4 Scattering of Intense Electromagnetic Radiation in Plasmas

Scattering of electromagnetic radiation in plasmas is a traditional area of study where the number of works is very large. Here we mention only several papers most closely related to the subject of this book.

Formalisms for treating laser-induced instabilities in plasmas were developed in [52, 53]. Raman scattering in tenuous plasmas and its effect on relativistic self-focusing were discussed in [54]. The production of high-energy electrons due to forward Raman scattering is simulated and studied experimentally in [55]. Time-asymptotic saturation of stimulated Raman scattering in homogeneous plasmas was considered in [56]. Raman backscattering in wake-field and beat-wave accelerators was investigated in [57]. The generation of backscattered harmonic radiation from interactions of intense laser radiation with beams and plasmas was also examined in [58]. A study of forward Raman scattering can be found in [59].

Instabilities of a circularly polarized monochromatic electromagnetic pump wave in plasma, which is described by an exact arbitrary amplitude solution of Maxwell and cold electron fluid dynamics equations [39], were treated in a number of works on electromagnetic radiation scattering at relativistic intensities. For slab geometry, the corresponding growth rates were studied numerically in [60] and estimated analytically in [61]. Paper [62] reports the study of the same growth rates and calculation of scattering angles in the framework of three-dimensional geometry under the assumption that the stimulated scattering harmonics obey the cold plasma electromagnetic wave dispersion relation. The same problem was treated using a special technique in [44,63] without the above assumption by calculating the instability growth rates numerically as eigenvalues of a matrix of an extremely high dimension. In particular, this approach made it possible to describe the generation of a continuum of scattered radiation. A similar study was performed in [64,65].

A more practically interesting case of linearly polarized relativistically intense electromagnetic radiation stimulated scattering presents a more complicated problem since here the ground state arbitrary amplitude solutions of plasma electrodynamics equations are not as readily available as in the case of circular polarization. An instability theory for an approximate monochromatic electromagnetic pump of linear polarization is developed in [66] assuming that the Raman scattering harmonics obey the electromagnetic wave dispersion relation. A more general stimulated scattering theory is proposed in [67, 68] where self-modulated solutions of the Akhiezer–Polovin problem [43] are used as ground states and the corresponding instability growth rates are computed numerically without any additional assumptions. An experimental observation of Raman scattering at near-relativistic intensities is reported in [69].

1.2.5 Interactions of Intense, Ultrashort, Laser Pulses with Underdense Plasmas

Theoretical investigation of the interactions of relativistically intense laser radiation and matter were stimulated by the achievement of relativistic intensities in experiments under conditions of tight focusing [70–73]. Numerous theoretical and experimental studies of nonlinear optical field-induced ionization resulting in plasma formation are available (for example, see [74–76]).

One of the first works on radiation propagation influenced by relativistic nonlinearity was published by Claire Ellen Max and co-workers [77]. She also examined the possibility of charge-displacement self-focusing [78]. An analytical estimate of the relativistic self-focusing threshold was given in [79]. Later, the problem involving a nonlinearity of the relativistic type was treated in [80].

Paper [81] constituted a milestone in the progress of the theory of interactions of relativistically strong laser radiation with plasmas. In this work, a laser beam propagation model embodied in the Schroedinger equation with relativistic and charge-displacement nonlinearity was developed, and its lowest axially symmetrical eigenmode was calculated. Furthermore, in this paper, under certain restrictions, the stability of the latter eigenmode in the framework of axially symmetrical geometry was inferred on the basis of numerical modeling. The Schroedinger equation with relativistic and charge-displacement nonlinearity remains one of the basic models of laser–plasma interactions at relativistic intensities. An important contribution to the investigation of this model was made in [82]: the aforementioned beam self-channeling condition according to which the nonlinear Schroedinger equation Hamiltonian must be negative was expanded to the case of relativistic and charge-displacement nonlinearity.

A large series of studies of relativistic intensity laser radiation and matter interaction was carried out by P. Sprangle and co-workers. They considered relativistic optical guiding, plasma wake-field and charged-particle acceleration, frequency shifts of laser pulses propagating in plasmas, stimulated backscattered harmonics excitation, etc. [58,83–88].

Slab geometry relativistically intense solitons propagating in cold underdense plasmas were found analytically in [89].

A concept of charged particle acceleration by high-frequency fields was proposed as early as 1958 [90,91]. At present, this is an area of vigorous research [92]. Particle-in-cell simulations of laser radiation self-focusing in plasmas are reported in [93].

A large number of papers was dedicated to relativistic and charge-displacement self-channeling of relativistically intense, ultrashort, laser pulses in cold underdense plasmas [7,8,94–97]. As a preliminary study, the simplified purely relativistic self-focusing problem (neglecting charge-displacement) was treated in [94], and the possibility of optical guiding was established. Relativistic and charge-displacement self-channeling was predicted theoretically in [7]. A countable set of axially symmetrical eigenmodes of the Schroedinger equation with relativistic and charge-displacement nonlinearity and the results of numerical simulations of the relativistic and charge-displacement self-channeling were presented in [7, 95]. As shown in these papers, laser pulses with powers greater than the relativistic and charge-displacement self-focusing critical power (also calculated in these works) propagate in guided regimes, and powerful laser amplitude distributions deep in the irradiated plasmas are given by the above lowest eigenmodes. The character of the relativistic and charge-displacement self-channeling was studied in finer detail numerically in [96]. Paper [8] reported the first experimental observation of relativistic and charge-displacement self-channeling. Filamentation of self-channeled relativistically intense laser pulses due to axial symmetry violations and related stability issues were considered in [97], where a concept of providing for self-channeling stability to azimuthal perturbation (and, consequently, for reliable concentration of laser beam energy in matter) by adjusting the incident radiation parameters in accord with a special theoretically developed algorithm was proposed (readers can find another numerical study of relativistic and charge-displacement filamentation in [98]). The application of this concept made it possible to increase the relativistic and charge-displacement distances in experiments [99].

Later, the theoretical results of [7] were verified in [35].

Another aspect of relativistic and charge-displacement self-channeling was investigated in [100]. Simulations using the same nonlinear Schroedinger equation model showed a giant broadening of the spectra of superintense laser pulses propagating in self-channeling regimes. Numerical studies reported in [101, 102] demonstrated a transformation of laser radiation in plasma into a two-dimensional solitary wave. An extensive numerical study of relativistic and charge-displacement self-channeling temporal dynamics (also in the framework of the nonlinear Schroedinger equation model) was presented in [103], where readers can also find a general theory of the self-channeling effect.

A series of interesting studies on interactions of relativistically intense, ultrashort, laser pulses and plasmas was performed by S.V. Bulanov and co-workers [104–112]. In particular, the possibility of focusing and defocusing laser radiation by plasma wakes was dealt with in [105]. In [106], a one-dimensional model of interactions of ultrashort, large aperture laser pulses with cold underdense plasmas was developed and used to investigate energy transfer from the optical field to plasma wakes. The acceleration of charged particles and photons in a narrow channel that emerges in plasma irradiated by a powerful laser is considered, and an increase in charged-particle energies and photon frequencies is calculated in [107]. Available papers on nonrelativistic wake-field excitation should also be mentioned [113, 114].

A novel approach to the theory of relativistic and charge-displacement self-channeling was implemented in [115–118], where the model of optical field evolution was based on the wave equation with relativistic and charge-displacement nonlinearity instead of the traditional Schroedinger equation. This provides for describing the impact of high order dispersion on the general propagation character. According to the simulation results reported in [115], the propagation picture following the first focus exhibits features different from those predicted with the help of the nonlinear Schroedinger equation. Instability of a monochromatic optical field was treated using the wave equation with relativistic and charge-displacement nonlinearity in [118]. Again, the growth rate distribution differed from that calculated on the basis of the nonlinear Schroedinger equation model.

The study reflected in [119] was another important step in the development of models of laser–plasma interactions at relativistic intensities. The three-dimensional model proposed in this work incorporates the generation of plasma waves by propagating electromagnetic radiation as well as relativistic and charge-displacement nonlinearities. This model was improved to provide for its adequate conservative properties, and the linear analysis of the instability of a relativistically strong monochromatic electromagnetic field corresponding to the combination of laser light filamentation and self-modulation was carried out in [120].

Synthesis of the model derived in [120] and the wave equation model of [115, 116] was carried out in [121], where additional terms responsible for optical field high-order dispersion and plasma inertia were added to the model established in [120], and their influence on the character of the electromagnetic field instability was examined.

A large number of works on prospective applications of relativistic and charge-displacement self-channeling are available (for example, see [122–124]). The idea of these works is to use relativistic and charge-displacement self-channeling as a method of concentrating energy in matter; the objective is to develop an X-ray laser. Readers can find an extensive survey of potential relativistic and charge-displacement self-channeling applications in [125].

Mechanisms of plasma nonlinearity other than plasma relativity, charge displacement, and plasmon excitation have been also studied. For example, frequency upshifting due to gas ionization and interactions with ionization fronts are studied in [126, 127], and ionization resulting in defocusing of laser radiation is treated in [128, 129].

1.2.6 Interactions of Relativistically Intense Laser Radiation with Overdense Plasmas

A kinetic theory of the relativistic skin effect that occurs when a high-intensity, ultrashort, laser pulse interacts with a solid target is developed in [130].

Spatially localized standing wave and soliton solutions for the relativistically intense electromagnetic field in an overdense plasma are presented in [131] (note that a similar problem was examined in [132] and [133] where ponderomotive cavitation was also discussed). Paper [131] also reports PIC simulations of soliton propagation and of breaking of such soliton solutions with overcritical amplitudes in overdense plasmas.

Numerical simulations revealed the existence of extreme self-generated magnetic fields in plasmas irradiated by powerful lasers [134]. An analytical theory of this phenomenon was developed in [135] with the help of the multiple timescale method.

Paper [108] was also dedicated to analytical investigation and PIC modeling of interactions of intense laser radiation with overdense plasmas. Laser radiation reflection from the plasma boundary, radiation absorption, harmonics generation, and low-frequency plasma oscillation excitation were discussed in this work. Vacuum heating of electrons in a laser plasma channel is studied in [109] as well as harmonics excitation and generation of quasi-static magnetic fields which, as it is shown, depend on laser radiation polarization. A PIC study of electron fluid vortices and magnetic wakes produced in plasmas by intense laser pulses undergoing the relativistic self-focusing is presented in [112].

2. Fundamentals of Cold Plasma Electrodynamics

This Chapter is dedicated to a brief outline of cold plasma relativistic fluid dynamics which typically serves as the basis for developing models of short powerful laser pulse propagation in matter. Following [32] and [103], we formulate a set of relativistic equations describing the dynamics of the electromagnetic field and plasma electron and ion components and the corresponding conservation laws. These equations are presented both in relativistic notation and in spatially three-dimensional form. We also consider the decomposition of plasma momentum into potential and curl parts. This decomposition simplifies the derivation of equations describing laser plasma interactions. The equations presented in this Chapter are used throughout this book.

2.1 Basic Cold Plasma Electrodynamics Equations in Relativistic Notation

Since rapid nonlinear ionization occurs at the front of a relativistically intense short laser pulse focused on matter, the pulse's central temporal part interacts with the resulting plasma. Experiments with subpicosecond excimer and Nd-glass laser pulses show that no substantial increase in plasma temperatures occurs during the pulse and that the plasma electrons are not heated up to temperatures substantially exceeding several keV [6]. At the same time, oscillation energies of electrons driven by the electromagnetic fields of relativistically intense laser pulses reach tens or hundreds of keV. Consequently, plasmas irradiated by short, powerful laser pulses can be considered cold.

Collisional and collisionless plasma heating is possible. Plasma collisional heating is inefficient for the following reasons. First, as is well known, the cross section of the elastic scattering of electrons by ions is a decreasing function of electron momentum [136]. This means that at high laser radiation intensities, the efficiency of electron gas collisional heating due to elastic scattering of electrons exhibiting ordered oscillations in a powerful electromagnetic field and colliding with ions is low. This fact is discussed in finer detail in [50]. Secondly, the potential for collisional plasma heating by short laser pulses tends to be limited due to the short interaction time.

Naturally, at higher intensities, a more important role is played by bremsstrahlung and above-the-threshold multiphoton absorption at the time of ionization [74, 137]. Both of these processes result in some additional heating of the electron gas.

Collisionless heating of electrons is much more significant. Electrons are accelerated by the ponderomotive force that expels them from the spatial area where the electromagnetic field intensity is high and by the electrostatic fields resulting from the generation of plasma waves. Charge separation behind the propagating laser pulse leads to ion acceleration (Coulomb explosion). Mechanisms of collisionless acceleration of charged particles and plasma heating were studied, for example, in [106, 108, 109, 138].

Below, we assume that the average kinetic energy of electron oscillations is much greater than their thermal energy, i.e.,

$$W_{\mathrm{osc}} = mc^2(\gamma - 1) \gg 3T/2\,,$$

where $\gamma = \sqrt{1 + I/I_{\mathrm{r}}}$ is the relativistic mass factor, I is the laser radiation intensity, and I_{r} is the relativistic intensity defined by (1.1). The above inequality can be rewritten as a criterion for the laser radiation intensity's magnitude:

$$I \gg 2 \cdot 10^{16}\, T/\lambda^2\, (\mathrm{W/cm^2})\,.$$

Here the temperature and the wavelength are measured in keV and microns, respectively. For a 1-keV temperature, the laser radiation intensity must be much greater than $2 \times 10^{16}\,\mathrm{W/cm^2}$ for Nd-glass lasers and $10^{17}\,\mathrm{W/cm^2}$ for excimer lasers. When this criterion is met it can be assumed that the plasma is cold and is driven only by the Lorentz force, whereas the impact of fluid dynamics pressure and the processes resulting in the emergence of this pressure is negligible.

As discussed in the introductory chapter of this book, it is necessary to account for the relativistic increase in the masses of charged particles in superintense optical fields. Primarily, this pertains to such light particles as electrons. Thus, the character of the nonlinear propagation of powerful ultrashort laser pulses into plasmas has to be described on the basis of relativistic cold plasma electrodynamics.

Consider a plasma consisting of electrons with charge $q = -e$, rest mass m, and ions with charge $Q = Ze$, rest mass M (in the notation adopted throughout this chapter $e > 0$). We assume that the plasma is quasi-neutral in the absence of exterior fields. Let n, $\boldsymbol{u}$, $\boldsymbol{p}$, $\boldsymbol{j}$ denote the electron concentration, velocity, momentum, and current. The corresponding quantities for ions are denoted by N, $\boldsymbol{U}$, $\boldsymbol{P}$, $\boldsymbol{J}$. The electron and ion four-currents are defined as

$$u^i = \left(\frac{c}{\sqrt{1 - u^2/c^2}}, \frac{\boldsymbol{u}}{\sqrt{1 - u^2/c^2}} \right), \tag{2.1}$$

$$U^i = \left(\frac{c}{\sqrt{1 - U^2/c^2}}, \frac{\boldsymbol{U}}{\sqrt{1 - U^2/c^2}} \right), \tag{2.2}$$

$$j^i = (c\rho_{\mathrm{e}}, \boldsymbol{j}) = (cqn, qn\boldsymbol{u})\,, \tag{2.3}$$

$$J^i = (c\rho_{\mathrm{i}}, \boldsymbol{J}) = (cQN, QN\boldsymbol{U}), \tag{2.4}$$

$$A^i = (\varphi, \boldsymbol{A}). \tag{2.5}$$

The electromagnetic field tensor is given by

$$F^{ik} = \frac{\partial A^k}{\partial x_i} - \frac{\partial A^i}{\partial x_k}. \tag{2.6}$$

Its components are expressed as

$$||F_{ik}|| = \left\| \begin{array}{cccc} 0 & E_x & E_y & E_z \\ -E_x & 0 & -B_z & B_y \\ -E_y & B_z & 0 & -B_x \\ -E_z & -B_y & B_x & 0 \end{array} \right\|, \tag{2.7}$$

where

$$\boldsymbol{E} = -c^{-1}\boldsymbol{A}_t - \nabla\varphi, \qquad \boldsymbol{B} = [\nabla, \boldsymbol{A}]. \tag{2.8}$$

The set of equations describing the dynamics of the system comprising the electromagnetic field and plasma is

$$m \frac{\partial u_k}{\partial x^l} u^l = \frac{q}{c} F_{kl} u^l, \tag{2.9}$$

$$M \frac{\partial U_k}{\partial x^l} U^l = \frac{Q}{c} F_{kl} U^l, \tag{2.10}$$

$$\frac{\partial F_{ik}}{\partial x^l} + \frac{\partial F_{kl}}{\partial x^i} + \frac{\partial F_{li}}{\partial x^k} = 0, \tag{2.11}$$

$$\frac{\partial F_{ik}}{\partial x_k} = -\frac{4\pi}{c}(j_i + J_i), \tag{2.12}$$

$$\frac{\partial j_i}{\partial x_i} = \frac{\partial J_i}{\partial x_i} = 0. \tag{2.13}$$

Equations (2.9) and (2.10) describe the motions of the plasma electron and ion fluids in the F_{ik} field. Equations (2.11) and (2.12) are the Maxwell equations in relativistic notation. Equations (2.13) are the electron and ion continuity equations.

The energy-momentum tensor for (2.9)–(2.13) is expressed as

$$T^{ik} = \frac{m}{q} u^i j^k + \frac{M}{Q} U^i J^k + \frac{1}{4\pi}\left(F^{il} F_l^k + \frac{1}{4} g^{ik} F_{lm} F^{lm}\right); \tag{2.14}$$

the corresponding conservation law is

$$\frac{\partial T^{ik}}{\partial x^k} = 0, \tag{2.15}$$

where g^{ik} is the metric tensor. To prove this, we calculate the divergence of the first two energy-momentum tensor terms

$$\frac{\partial}{\partial x^k}\left(\frac{m}{q}\,u^i j^k + \frac{M}{Q}\,U^i J^k\right) = \frac{m}{q}\,\frac{\partial u^i}{\partial x^k}\,j^k + \frac{M}{Q}\,\frac{\partial U^i}{\partial x^k}\,J^k$$
$$= \frac{1}{c}\,F^{ik}(j_k + J_k)\,. \tag{2.16}$$

It follows from the continuity equations (2.13) that the terms proportional to $\partial j^k/\partial x^k$ and $\partial J^k/\partial x^k$ in the divergence of products $u^i j^k$ and $U^i J^k$ are equal to zero. Equations of motion for electrons and ions can be cast in the form

$$m\,\frac{\partial u_k}{\partial x^l}\,j^l = \frac{q}{c}\,F_{kl}j^l\,, \qquad M\,\frac{\partial U_k}{\partial x^l}\,J^l = \frac{Q}{c}\,F_{kl}J^l\,. \tag{2.17}$$

Substituting the four-current from (2.12) in (2.16),

$$\frac{1}{c}\,F^{ik}(j_k + J_k) = -F^{ik}\,\frac{1}{4\pi}\,\frac{\partial F_{kl}}{\partial x_l}$$
$$= -\frac{1}{4\pi}\left[\frac{\partial}{\partial x_l}(F^{ik}F_{kl}) - F_{kl}\frac{\partial}{\partial x_l}\,F^{ik}\right]. \tag{2.18}$$

The last term of this equation can be transformed as follows:

$$F_{kl}\frac{\partial}{\partial x_l}\,F^{ik} = \frac{1}{2}\,F_{kl}\left(\frac{\partial}{\partial x_l}\,F^{ik} + \frac{\partial}{\partial x_k}\,F^{li}\right) = -\frac{1}{2}\,F_{kl}\frac{\partial}{\partial x_i}\,F^{kl}$$
$$= -\frac{1}{4}\,\frac{\partial}{\partial x_i}\,F_{kl}F^{kl} = -\frac{1}{4}\,\frac{\partial}{\partial x_k}\,g^{ik}F_{lm}F^{lm}\,. \tag{2.19}$$

Here, we used (2.11). Finally,

$$\frac{1}{c}\,F^{ik}(j_k + J_k) = -\frac{\partial}{\partial x^k}\,\frac{1}{4\pi}\left(F^{il}F_l^k + \frac{1}{4}\,g^{ik}F_{lm}F^{lm}\right). \tag{2.20}$$

Combining (2.20) and (2.16), we arrive at (2.15).

Equations (2.9)–(2.13) and the corresponding conservation law (2.15) involving the energy-momentum tensor (2.14) components are the basic equations of cold relativistic plasma electrodynamics. In this book, we will proceed from this set of equations.

2.2 Basic Equations in 3-D Form

System (2.12) comprises four equations for an electromagnetic field. Equation (2.11) is satisfied identically, which can be verified directly with the help of (2.6). However the physical fields have only three components (for example, this applies to $\boldsymbol{E}$, whereas $\boldsymbol{B}$ can be calculated when the electric field is known). Therefore, an additional gauge condition can be imposed on the potentials. It follows from (2.12) and (2.6) that

$$\frac{\partial^2 A_i}{\partial x_k \partial x^k} - \frac{\partial^2 A_k}{\partial x^i \partial x_k} = \frac{4\pi}{c} (j_i + J_i) . \tag{2.21}$$

The relativistically invariant Lorenz condition for the vector potential is

$$\frac{\partial A_k}{\partial x_k} = 0 . \tag{2.22}$$

Maxwell equations with a Lorentz gauge are Lorentz-invariant.

Generally, the potentials can be subject to an arbitrary condition. However, in this case the equations including the gauge condition may have a different form in different reference frames. The only case where it is convenient to use noninvariant gauges is when only one reference frame is used.

Lorentz Gauge. Equations (2.9)–(2.13) in Lorentz gauge are

$$\boldsymbol{p}_t + (\boldsymbol{u}, \nabla)\boldsymbol{p} = q\Big(-c^{-1}\boldsymbol{A}_t - \nabla\varphi + c^{-1}\big[\mathbf{u}, [\nabla, \boldsymbol{A}]\big]\Big) , \tag{2.23}$$

$$\boldsymbol{P}_t + (\boldsymbol{U}, \nabla)\boldsymbol{P} = Q\Big(-c^{-1}\boldsymbol{A}_t - \nabla\varphi + c^{-1}\big[\mathbf{U}, [\nabla, \boldsymbol{A}]\big]\Big) , \tag{2.24}$$

$$(\triangle - c^{-2}\partial_t^2)\boldsymbol{A} = -\frac{4\pi}{c} (\boldsymbol{j} + \boldsymbol{J}) , \tag{2.25}$$

$$(\triangle - c^{-2}\partial_t^2)\varphi = -4\pi\rho , \tag{2.26}$$

$$c^{-1}\varphi_t + (\nabla, \boldsymbol{A}) = 0 , \tag{2.27}$$

$$\boldsymbol{j} + \boldsymbol{J} = qn\boldsymbol{u} + QN\boldsymbol{U} , \tag{2.28}$$

$$\rho = qn + QN , \tag{2.29}$$

$$\boldsymbol{u} = \frac{\boldsymbol{p}}{m\gamma_{\mathrm{e}}} , \qquad \gamma_{\mathrm{e}} = \sqrt{1 + \left(\frac{|\boldsymbol{p}|}{mc}\right)^2} , \tag{2.30}$$

$$\boldsymbol{U} = \frac{\boldsymbol{P}}{M\gamma_{\mathrm{i}}} , \qquad \gamma_{\mathrm{i}} = \sqrt{1 + \left(\frac{|\boldsymbol{P}|}{Mc}\right)^2} . \tag{2.31}$$

Here and everywhere below, the index t denotes the derivative in time. Equations (2.23)–(2.24) describe the electron and ion fluid dynamics of a cold plasma in an exterior electromagnetic field. Relations (2.25)–(2.26) are Maxwell equations for the vector and scalar potentials. Equation (2.27) represents the Lorentz gauge. Equations (2.28)–(2.29) define the plasma current and charge densities. Equations (2.30)–(2.31) are the relativistic relations between velocity and momentum.

The continuity equation corresponding to charge conservation follows from the above equations. Calculating the divergence of (2.25) and the derivative in time of (2.26), summing the resulting two equations, and using the gauge condition (2.27), we find, that

$$\rho_t + \Big(\nabla, (\boldsymbol{j} + \boldsymbol{J})\Big) = 0 . \tag{2.32}$$

Coulomb Gauge. In the spatially three-dimensional form (2.21) are

$$(\triangle - c^{-2}\partial_t^2)\boldsymbol{A} - \nabla\Big(c^{-1}\varphi_t + (\nabla, \boldsymbol{A})\Big) = -\frac{4\pi}{c}\,(\boldsymbol{j} + \boldsymbol{J})\,, \tag{2.33}$$

$$(\triangle - c^{-2}\partial_t^2)\varphi + c^{-1}\partial_t\Big(c^{-1}\varphi_t + (\nabla, \boldsymbol{A})\Big) = -4\pi\rho\,. \tag{2.34}$$

In the Coulomb gauge,

$$(\nabla, \boldsymbol{A}) = 0\,, \tag{2.35}$$

$$(\triangle - c^{-2}\partial_t^2)\boldsymbol{A} - c^{-1}\nabla\varphi_t = -\frac{4\pi}{c}\,(\boldsymbol{j} + \boldsymbol{J})\,, \tag{2.36}$$

$$\Delta\varphi = -4\pi\rho\,. \tag{2.37}$$

Gauge in which $\boldsymbol{A}_\parallel = 0$. In some problems of coherent radiation propagation, it is convenient to put Maxwell equations in the gauge where the component of the vector potential in the propagation direction equals zero. In this case, $A_\parallel = 0$, and

$$(\triangle - \partial_t^2)\boldsymbol{A}_\perp - \nabla_\perp\Big(c^{-1}\varphi_t + (\nabla_\perp, \boldsymbol{A}_\perp)\Big) = -\frac{4\pi}{c}\,(\mathbf{j}_\perp + \boldsymbol{J}_\perp)\,, \tag{2.38}$$

$$\nabla_\parallel\Big(c^{-1}\varphi_t + (\nabla_\perp, \boldsymbol{A}_\perp)\Big) = -\frac{4\pi}{c}\,(\mathbf{j}_\parallel + \boldsymbol{J}_\parallel)\,, \tag{2.39}$$

$$\Delta\varphi + c^{-1}(\nabla_\perp, \boldsymbol{A}_{\perp_t}) = -4\pi\rho\,. \tag{2.40}$$

Energy and Momentum Conservation. The energy and momentum conservation laws for (2.23)–(2.31) are expressed in four-dimensional form by (2.15). The energy-momentum tensor (2.14) components are given by

$$T^{00} = \frac{nmc^2}{\sqrt{1-u^2/c^2}} + \frac{NMc^2}{\sqrt{1-U^2/c^2}} + \frac{1}{8\pi}(|\boldsymbol{E}|^2 + |\boldsymbol{B}|^2)\,, \tag{2.41}$$

$$T^{0\alpha} = \frac{nmcu_\alpha}{\sqrt{1-u^2/c^2}} + \frac{NMcU_\alpha}{\sqrt{1-U^2/c^2}} + \frac{1}{4\pi}\,[\boldsymbol{E}, \boldsymbol{B}]_\alpha\,, \tag{2.42}$$

$$T^{\alpha\beta} = np_\alpha u_\beta + NP_\alpha U_\beta - \sigma_{\alpha\beta}\,, \tag{2.43}$$

$$\sigma_{\alpha\beta} = \frac{1}{4\pi}(E_\alpha E_\beta + B_\alpha B_\beta) - \frac{1}{8\pi}(|\boldsymbol{E}|^2 + |\boldsymbol{B}|^2)\delta_{\alpha\beta}\,. \tag{2.44}$$

In spatially three-dimensional form, the conservation laws are written as

$$\partial_t\Big[mc^2\gamma_{\mathrm{e}}n + Mc^2\gamma_{\mathrm{i}}N + \frac{1}{8\pi}(|\boldsymbol{E}|^2 + |\boldsymbol{B}|^2)\Big] + \Big(\nabla, mc^2\gamma_{\mathrm{e}}n\boldsymbol{u} + Mc^2\gamma_{\mathrm{i}}N\boldsymbol{U} + \frac{1}{4\pi}\,[\boldsymbol{E}, \boldsymbol{B}]\Big) = 0\,, \tag{2.45}$$

$$\partial_t\Big(np_\alpha + NP_\alpha + \frac{1}{4\pi c}[\boldsymbol{E},\boldsymbol{B}]_\alpha\Big) + \frac{\partial}{\partial x_\beta}\Big[(np_\alpha u_\beta + NP_\alpha U_\beta) + \frac{1}{8\pi}(|\boldsymbol{E}|^2 + |\boldsymbol{B}|^2)\delta_{\alpha\beta} - \frac{1}{4\pi}(E_\alpha E_\beta + B_\alpha B_\beta)\Big] = 0\,. \tag{2.46}$$

Here, $(1/8\pi)(|\boldsymbol{E}|^2 + |\boldsymbol{B}|^2)$ is the electromagnetic field energy density,$(1/4\pi)$ $\times[\boldsymbol{E},\boldsymbol{B}]$ is the energy flux density, and $\sigma_{\alpha\beta}$ is the field momentum flux density. The electric and magnetic fields are related to the vector and scalar potentials by (2.8).

2.3 Potential and Vortex Components of Momentum

The term $(\boldsymbol{u},\nabla)\boldsymbol{p}$ in (2.23) can be transformed in the following way:

$$\begin{aligned}(\boldsymbol{u},\nabla)\boldsymbol{p} &= (m\gamma_e)^{-1}(\boldsymbol{p},\nabla)\boldsymbol{p} = (m\gamma_e)^{-1}\Big(\frac{1}{2}\nabla|\boldsymbol{p}|^2 - [\boldsymbol{p},[\nabla,\boldsymbol{p}]]\Big)\\ &= mc^2\nabla\gamma_e - [\boldsymbol{u},[\nabla,\boldsymbol{p}]]\,.\end{aligned}$$

Then, (2.23) becomes

$$\partial_t\left(\boldsymbol{p} + \frac{q}{c}\boldsymbol{A}\right) - \left[\mathbf{u},\left[\nabla,\left(\boldsymbol{p} + \frac{q}{c}\boldsymbol{A}\right)\right]\right] = -q\nabla\varphi - mc^2\nabla\gamma_e\,. \tag{2.47}$$

We seek a solution for the momentum as a sum of three terms

$$\boldsymbol{p} = -\frac{q}{c}\boldsymbol{A} + \boldsymbol{p}_0 + \boldsymbol{p}_1\,, \tag{2.48}$$

where $\boldsymbol{p}_0$ is a potential vector and $\boldsymbol{p}_1$ is a curl vector, namely,

$$[\nabla,\boldsymbol{p}_0] = 0\,,\qquad (\nabla,\boldsymbol{p}_0)\neq 0,\qquad [\nabla,\boldsymbol{p}_1]\neq 0\,,\qquad (\nabla,\boldsymbol{p}_1) = 0.$$

Combining (2.48) and (2.47) and calculating the curl of the former equation's left- and right-hand sides, we arrive at an equation for $\boldsymbol{M} = [\nabla,\boldsymbol{p}_1]$:

$$\partial_t\boldsymbol{M} - [\nabla,[\boldsymbol{u},\boldsymbol{M}]] = 0\,. \tag{2.49}$$

Calculating the divergence of (2.47), we get an equation for $D = (\nabla,\boldsymbol{p}_0)$,

$$\partial_t D - (\nabla,[\boldsymbol{u},\boldsymbol{M}]) = -q\Delta\varphi - mc^2\Delta\gamma_e\,. \tag{2.50}$$

Obviously, (2.49) and (2.50) are nonlinear since there is a relativistic relation between $\boldsymbol{u}$ and $\boldsymbol{p}$ that involves both $\boldsymbol{p}_0$ and $\boldsymbol{p}_1$.

Substantial simplifications can stem from the fact that under certain circumstances there is no need to deal with the generation of plasma vortices given by $\boldsymbol{M}$. If a laser pulse interacts with a plasma and all of its motions at an initial time were potential, there will be no generation of curl motions in it. This follows from (2.49). Indeed, if $\boldsymbol{M} = 0$ and $\partial\boldsymbol{M}/\partial x_\alpha = 0$ at $t = 0$ everywhere in a plasma, then $\boldsymbol{M}_t = 0$, and therefore the generation of curls does not occur. Exactly this situation is considered in the following chapters of this book.

Under experimental conditions, vortex fluctuations may be present in plasmas. Even if the magnitude of $\boldsymbol{M}$ is small, it is possible that $\partial \boldsymbol{M}/\partial x_\alpha \neq 0$. The issue whether significant electron vortices evolving out of these fluctuations can influence the laser radiation dynamics substantially remains open to discussion. Electron vortices remaining behind propagating laser pulses can give rise to magnetic wakes forming complicated stable structures propagating in plasmas at relatively low velocities [111].

Below, we assume that $\boldsymbol{p}_1 \equiv 0$. In this case, the electron and ion equations of motion (2.23)–(2.24) are reduced to

$$\boldsymbol{p} = -\frac{q}{c}\boldsymbol{A} + \boldsymbol{p}_0 \,, \tag{2.51}$$

$$\boldsymbol{P} = -\frac{Q}{c}\boldsymbol{A} + \boldsymbol{P}_0 \,, \tag{2.52}$$

$$\partial_t \boldsymbol{p}_0 = -q\nabla\varphi - mc^2\nabla\gamma_{\mathrm{e}} \,, \tag{2.53}$$

$$\partial_t \boldsymbol{P}_0 = -Q\nabla\varphi - Mc^2\nabla\gamma_{\mathrm{i}} \,, \tag{2.54}$$

and the electron and ion currents are expressed as

$$\begin{aligned} \boldsymbol{j} + \boldsymbol{J} &= qn\boldsymbol{u} + QN\boldsymbol{U} = \frac{qn}{m\gamma_{\mathrm{e}}}\boldsymbol{p} + \frac{QN}{M\gamma_{\mathrm{i}}}\boldsymbol{P} \\ &= -\left(\frac{q^2 n}{mc\gamma_{\mathrm{e}}} + \frac{Q^2 N}{Mc\gamma_{\mathrm{i}}}\right)\boldsymbol{A} + \frac{qn}{m\gamma_{\mathrm{e}}}\boldsymbol{p}_0 + \frac{QN}{M\gamma_{\mathrm{i}}}\boldsymbol{P}_0 \,. \end{aligned} \tag{2.55}$$

Equations (2.51)–(2.55) combine better with Maxwell equations in the Coulomb gauge since in this case, the vector potential is a curl vector and (2.51)–(2.52) present a decomposition of the electron and ion momenta into potential and curl parts.

Thus, the basic cold relativistic plasma electrodynamics equations used for the development of models of powerful laser pulse propagation in matter are formulated in this chapter. We have considered them in both relativistic notation and in spatially three-dimensional form. Conservation laws for the electromagnetic field and plasma system energy and momentum are associated with these equations. Separation of the curl and potential components of the electron and ion momenta is also described.

2.4 Electron Fluid Dynamics with Inertially Frozen Ions

In interaction of ultrashort (femtosecond) laser pulses with plasmas, it is natural to assume that due to their relatively high inertia, ions are frozen in space and constitute an immobile positive background for the laser-driven motions of the electron fluid. In the framework of this approximation $\boldsymbol{P} = 0$, $\gamma_i = 1$, and $N = N_0 = \mathrm{const}$. For a fully ionized plasma which was quasi-neutral prior to the laser pulse incidence, the last expression becomes $ZN_0 = n_0$, where Z

is the ion charge number and n_0 is the unperturbed electron concentration. It is convenient to cast the electron fluid dynamics equations into dimensionless form as follows: the field vector and scalar potentials are normalized by mc^2/e; the electron momentum is normalized by mc; the electron concentration by its unperturbed value n_0; time by ω_p^{-1}; and the coordinates by c/ω_p, where ω_p is the unperturbed plasma electron frequency. The resulting set of equations is

$$(\triangle - \partial_t^2)\,\mathbf{A} = \nabla\varphi_t + \gamma^{-1} n\,\boldsymbol{p}\,, \tag{2.56}$$

$$\triangle\varphi = n - 1\,, \tag{2.57}$$

$$(\nabla, \mathbf{A}) = 0\,, \tag{2.58}$$

$$(\boldsymbol{p} - \boldsymbol{A})_t - \gamma^{-1}\,[\boldsymbol{p}, [\nabla, \boldsymbol{p} - \boldsymbol{A}]] = \nabla(\varphi - \gamma)\,, \tag{2.59}$$

$$\gamma = \sqrt{1 + |\boldsymbol{p}|^2}\,. \tag{2.60}$$

For brevity, we use the same symbols for the normalized values as for the corresponding dimensional ones. The gauge given by (2.58) is equivalent to the normalized continuity equation,

$$n_t + (\nabla, \frac{n}{\gamma}\boldsymbol{p}) = 0\,. \tag{2.61}$$

2.4.1 Canonical Variables

An alternative way to implement the decomposition of the electron fluid momentum into potential and nonpotential parts was proposed in [51, 139], where the three components of the momentum are related to three other scalars ψ, α, and β according to

$$\boldsymbol{p} = \boldsymbol{A} + \nabla\psi + \alpha\nabla\beta\,. \tag{2.62}$$

Three scalar equations governing the new fluid dynamics variables are derived from the Euler vector equation (2.59):

$$\psi_t = \varphi - \gamma + \alpha\gamma^{-1}(\boldsymbol{p}, \nabla\beta)\,, \tag{2.63}$$

$$\alpha_t + \gamma^{-1}(\boldsymbol{p}, \nabla\alpha) = 0, \qquad \beta_t + \gamma^{-1}(\boldsymbol{p}, \nabla\beta) = 0\,. \tag{2.64}$$

Here, ψ is easily identified as the potential of the field–plasma system generalized momentum, whereas α, and β are known in hydrodynamics as Clebsch variables. Obviously, vortex-free field–plasma systems are described by $\alpha = \beta = 0$. According to [51], the set of variables α/n, β, ψ, n, $\boldsymbol{A}$, and the appropriately normalized electric field $\boldsymbol{E}$ are the canonical variables of the plasma model considered.

2.4.2 Examples of Exact Solutions

Below, we consider two simple examples of solutions of the equations of plasma electron fluid dynamics. The first pertains to the case where the radiation frequency is greater than the plasma frequency (underdense plasma) and the plasma is transparent to the propagating electromagnetic wave. The second solution presented below describes a standing optical wave in an overdense plasma.

Circularly Polarized Monochromatic Electromagnetic Waves in Cold Underdense Plasmas. The simplest solution of (2.56)–(2.61) is the well known

$$\mathbf{A} = \frac{1}{2}(\mathbf{e}_1 + \mathrm{i}\mathbf{e}_2)\, a_0 \exp[\mathrm{i}(k_0 x_3 - \omega_0 t)] + c.c., \tag{2.65}$$

describing a plane, monochromatic, circularly polarized wave with a spatially uniform intensity distribution, and the dispersion relation

$$\omega_0^2 - k_0^2 = \gamma_0^{-1}, \tag{2.66}$$

where

$$\gamma_0 = \sqrt{1 + a_0^2} \tag{2.67}$$

and the corresponding fluid dynamics quantities are given by

$$\phi = \gamma_0\,, \qquad \boldsymbol{p} = \boldsymbol{A}\,, \qquad n = 1\,. \tag{2.68}$$

This is easy to verify directly. The simplicity of the above solution is due to the fact that the propagating electromagnetic field causes an increase in the electron mass density, whereas the plasma remains uniform. More general solutions will be presented in the next chapter (also, see [43, 44]).

As follows from the dispersion relation (2.66), the electromagnetic wave phase velocity (normalized by the speed of light) makes

$$q = \frac{\omega_0}{k_0} = \sqrt{1 + \frac{1}{\gamma_0 k_0^2}}\,. \tag{2.69}$$

Since $k_\mathrm{p}(a_0^2) = \gamma_0^{-1/2}$ is the plasma electron wave vector modified by the relativistic increase in electron mass caused by field-induced oscillations (and normalized by its unperturbed value), it follows obviously from (2.66) that the frequency of the electromagnetic wave corresponding to (2.65) is greater than the plasma frequency, i.e., the above solution describes propagation in underdense plasmas. Practically, the most interesting situation is the Compton limit where the electromagnetic wave frequency (and, consequently, wave number) is much greater than the plasma electron frequency. Equation (2.69) shows that in this case the electromagnetic wave phase velocity $q = \sqrt{1 + (k_\mathrm{p}/k_0)^2}$ exceeds the speed of light by a small value. In the next chapter, an expansion in the small parameter $\epsilon = k_\mathrm{p}/k_0$ related to this value will make it possible to develop asymptotic solutions to the Maxwell and electron fluid dynamics equations describing a more general class of nonlinear electromagnetic waves in plasmas.

Note that despite its elementary character, the solution of the Maxwell and electron fluid dynamics equations presented in this section describes the propagation of an electromagnetic wave having an arbitrarily high amplitude into a cold underdense plasma. Thanks to the latter circumstance, this solution finds various applications in the theory of the interactions of relativistically intense laser pulses with matter. First of all, its instabilities serve as models of the stimulated scattering of laser radiation by plasmons. This topic is the subject of Chap. 4 of this book. Second, slow dependence on time and coordinates can be introduced into the above plane wave amplitude in the framework of the envelope approximation, which makes it possible to develop models of interactions between plasmas and laser pulses with spatially localized intensity distributions. This topic will be discussed later in Chap. 6.

Standing Waves and Solitons in Overdense Plasmas. A more complicated class of solutions to the Maxwell and electron fluid dynamics equations was developed in [131–133]. These solutions describe electromagnetic waves with spatially localized amplitudes in overdense plasmas.

Consider a vortex-free field–plasma system in the slab geometry. Following [131–133], we seek its solutions using the ansatz

$$\boldsymbol{A} = (0, a_0(x)\cos(st), a_0(x)\sin(st)) \ ,$$

$$\phi = \phi_0(x) \,, \qquad n = n_0(x) \,, \qquad \psi = \psi_0(x) \,.$$

The boundary conditions for spatially localized solutions are

$$a_0(\pm\infty) = \psi_{0_x}(\pm\infty) = 0 \,, \qquad \phi_0(\pm\infty) = n_0(\pm\infty) = 1 \,. \tag{2.70}$$

In this case, (2.56) is reduced to a scalar equation for the field amplitude:

$$\left(s^2 + \partial_x^2\right) a_0 = \frac{n_0}{\gamma} a_0 \,. \tag{2.71}$$

The nonlinearity of the above equation can be expressed as a function of a_0 and its derivatives. From the continuity equation (2.61), $\psi_{0_x} \equiv 0$, so that

$$\gamma = \sqrt{1 + |a_0(x)|^2} \,. \tag{2.72}$$

Next, combining (2.57) and (2.63),

$$n_0 = 1 + \gamma_{xx} \,. \tag{2.73}$$

The first integral of (2.71)–(2.73) is

$$a_{0_x}^2 + s^2 a_0^2 - 2\gamma - \gamma_x^2 = 2E \,, \tag{2.74}$$

and from boundary conditions (2.70), $E = -1$. The change of variables $a_0(x) = \sinh u(x)$ and $\gamma = \cosh u(x)$ makes it possible to integrate (2.74); the result is

$$a_0(x) = \frac{2\sqrt{1-s^2}\cosh(\sqrt{1-s^2}\,x)}{\cosh^2(\sqrt{1-s^2}\,x) + s^2 - 1}\,. \tag{2.75}$$

The relativistic mass factor, the normalized electric and magnetic fields, the electron momentum and concentration are expressed as [131]

$$\gamma = \frac{\cosh^2(\sqrt{1-s^2}\,x) - s^2 + 1}{\cosh^2(\sqrt{1-s^2}\,x) + s^2 - 1}\,,$$

$$E_1 = \frac{2(1-s^2)^{3/2}\sinh(2\sqrt{1-s^2}\,x)}{\left(\cosh^2(\sqrt{1-s^2}\,x) + s^2 - 1\right)^2}\,,$$

$$E_2 = \frac{2\sqrt{1-s^2}\cosh(\sqrt{1-s^2}\,x)}{\cosh^2(\sqrt{1-s^2}\,x) + s^2 - 1} s\sin(st)\,,$$

$$E_3 = -\frac{2\sqrt{1-s^2}\cosh(\sqrt{1-s^2}\,x)}{\cosh^2(\sqrt{1-s^2}\,x) + s^2 - 1} s\cos(st)\,,$$

$$\boldsymbol{B} = \frac{2(1-s^2)\sinh(\sqrt{1-s^2}\,x)\left(\cosh^2(\sqrt{1-s^2}\,x) - s^2 + 1\right)}{\left(\cosh^2(\sqrt{1-s^2}\,x) + s^2 - 1\right)^2} \times (0, \sin(st), -\cos(st))\,,$$

$$\boldsymbol{p} = \boldsymbol{A} = \frac{2\sqrt{1-s^2}\cosh(\sqrt{1-s^2}\,x)}{\cosh^2(\sqrt{1-s^2}\,x) + s^2 - 1}\,(0, \cos(st), \sin(st))\,,$$

$$n = 1 + (1-s^2)2\frac{\cosh(4\sqrt{1-s^2}\,x) + 2(1-2s^2)\cosh(2\sqrt{1-s^2}) - 3}{\left(\cosh^2(\sqrt{1-s^2}\,x) + s^2 - 1\right)^3}\,.$$

Naturally, the above solution exists only for $s^2 < 1$, and since s is the normalized electromagnetic wave frequency and the unperturbed plasma frequency is equal to unity in the normalization we use, (2.75) describes the propagation of an electromagnetic field in an overdense plasma.

An obvious limitation of the above derivation is that the expression for the electron density must yield nonnegative values. The electron concentration can equal zero in a certain spatial domain. This possibility was carefully examined in [132, 133]. In particular Marburger and Tooper estimated the threshold electromagnetic field at which electron cavitation occurs [132] (the corresponding condition is easy to obtain from (2.57)) by matching logarithmic derivatives of solutions at the cavitation boundary with those for the spatial domain which is free of electrons. The potential impact of the inhomogeneity of the ion distribution on the electromagnetic waves in plasmas was also discussed in these papers.

Bulanov and co-workers considered a situation more general than that embodied in the initial assumptions of this section; namely instead of standing waves, they treated solitons moving in plasma at nonrelativistic velocities. Such electromagnetic relativistically intense solitons were observed in both underdense and overdense plasmas in PIC simulations [106, 108]. According to [131], a critical value of the relativistic soliton amplitude exists at which the electron longitudinal momentum (which, in contrast to the case of standing waves, is not zero for moving solitons) becomes singular, and soliton breaking occurs. Heating and acceleration of electrons up to the energies of 3–5 mc^2 was predicted in [131] for overcritical soliton amplitudes.

3. Relativistically Intense Electromagnetic Waves in Plasmas

As we saw in the previous chapter, if a plasma is strongly underdense, the phase velocity of an intense electromagnetic wave propagating in it is just slightly greater than the speed of light. Though this fact was demonstrated in the framework of an elementary example of a circularly polarized monochromatic electromagnetic wave which propagates in a plasma without perturbing its electron concentration or exciting any longitudinal electron motions and fields, the same statement remains valid for a much broader range of physical situations. In this chapter, we will introduce a small parameter characterizing the proximity of the electromagnetic radiation phase velocity in an underdense plasma and the speed of light and derive the corresponding asymptotic solutions describing the propagation of nonlinear relativistically intense waveforms comprising an electromagnetic and a plasma wave; their coupling results in nonlinear amplitude and phase self-modulation.

The most general description of electromagnetic waves in relativistic plasmas is embodied in the classic Akhiezer–Polovin problem [39]. In its unnormalized form, this problem follows from the Maxwell and electron fluid dynamics equations when the electromagnetic field vector and scalar potentials as well as the electron concentration and momentum depend on a single phase variable $kx - \omega t$. In the case outlined above, it becomes possible to separate the fast phase of the electromagnetic field and that of the slow plasmon and establish an averaged equation governing the laser–plasma interaction. As we shall see below, this approach applies to various electromagnetic radiation polarizations and makes it possible to derive approximate solutions for the field and plasma quantities; relations between the local radiation amplitude and frequency shifts; and an expression relating the average electromagnetic field intensity, its average wave number, and the field–plasma coupling parameter.

3.1 The Akhiezer–Polovin Problem

Once again, let us return to the system of Maxwell equations (Coulomb gauge) and the equations of electron fluid dynamics (2.56)–(2.61) in slab geometry: $x_1 = x_2 = 0$. Following [39], we seek its solutions depending on a single variable $\xi = x_3 - qt$, where $q = \sqrt{1+\epsilon^2}$ is the phase velocity and ϵ is a parameter (since, obviously, $q > 1$, this expression for the phase velocity automatically implies that the electromagnetic radiation phase velocity in unmagnetized plasmas is greater than the speed of light). The result of transition to the above variable is a set of coupled ordinary differential equations:

$$\epsilon^2 \boldsymbol{A}_{\xi\xi} + \frac{n}{\gamma}\boldsymbol{A} = 0\,, \tag{3.1}$$

$$\varphi_{\xi\xi} = n - 1\,, \tag{3.2}$$

where

$$n = \frac{\sqrt{1+\epsilon^2}\,\gamma}{\sqrt{1+\epsilon^2}\,\gamma - p_3}\,, \tag{3.3}$$

$$\sqrt{1+\epsilon^2}\,p_3 = \gamma - \varphi\,, \tag{3.4}$$

$$\gamma = \sqrt{1+|\boldsymbol{A}|^2+p_3^2} \tag{3.5}$$

(integrating the continuity equation, we assumed that $\varphi = 1$ for the unperturbed state of the field–plasma system). The longitudinal component of the electron fluid momentum and the electron concentration are expressed as functions of $\boldsymbol{A}$ and ϕ using (3.3)–(3.5):

$$p_3 = \epsilon^{-2}\left(\sqrt{\varphi^2+\epsilon^2\,(1+|\boldsymbol{A}|^2)} - \sqrt{1+\epsilon^2}\,\varphi\right)\,, \tag{3.6}$$

$$n - 1 = \epsilon^{-2}\left[1 - \sqrt{\frac{1+\epsilon^2}{\varphi^2+\epsilon^2\,(1+|\boldsymbol{A}|^2)}}\,\varphi\right]\,. \tag{3.7}$$

Substituting these formulas in (3.1) and (3.2), we arrive at the classic Akhiezer–Polovin problem [39]:

$$\epsilon^2\,\boldsymbol{A}_{\xi\xi} + F(\mathbf{A},\varphi,\epsilon)\,\mathbf{A} = 0\,, \tag{3.8}$$

$$\epsilon^2\varphi_{\xi\xi} + F(\mathbf{A},\varphi,\epsilon)\,\varphi - 1 = 0\,; \tag{3.9}$$

the nonlinearity is

$$F(\mathbf{A},\varphi,\epsilon) = \sqrt{\frac{1+\epsilon^2}{\varphi^2+\epsilon^2\,(1+|\boldsymbol{A}|^2)}}\,. \tag{3.10}$$

The invariants for the above equations are given by [39]

$$\frac{1}{2}(\epsilon^2|\boldsymbol{A}_\xi|^2+\varphi_\xi^2) + \epsilon^{-2}\left[\sqrt{(1+\epsilon^2)\,[\varphi^2+\epsilon^2(1+|\boldsymbol{A}|^2)]} - \varphi\right] = \text{const}\,,$$

$$A_1 A_{2_\xi} - A_2 A_{1_\xi} = \text{const}\,.$$

The Akhiezer–Polovin problem describes the nonlinear propagation of an intense electromagnetic field into a cold plasma, excitation of a longitudinal field and of plasma oscillations by the transversely polarized radiation, and the corresponding electromagnetic field self-effect. Just like (2.56)–(2.61), equations comprising problem (3.8)–(3.10) are fully relativistic and pertain to electromagnetic waves with arbitrarily high amplitudes.

3.2 Linearly Polarized Plane Electromagnetic Waves

In this chapter, our attention will be focused primarily on the most practically significant case of linear polarization of laser radiation propagating in a plasma (results concerning circular polarization will also be discussed briefly). Obviously, (3.8)–(3.10) admit the corresponding solutions, namely those where $A_2 \equiv 0$. In this case, $M = 0$ in the conservation law (3.11).

3.2.1 Self-Modulation at Relativistic Intensities

Nonlinear amplitude and phase modulation of relativistically intense electromagnetic waves propagating in cold underdense plasmas with phase velocities close to the speed of light was studied in the framework of the Akhiezer – Polovin problem in [43, 44]. A numerical solution of (3.8)–(3.10) is depicted in Fig. 3.1 (linear polarization, $A_2 \equiv 0$). This solution corresponds to the following (artificial) initial conditions: $A_1(0) = 1.2$, $A_{1_\xi}(0) = 5$, $\varphi(0) = 2$, $\varphi_\xi(0) = 0$, and the problem parameter is taken as $\epsilon = 0.1$. Additional examples of numerical solutions of the Akhiezer–Polovin problem are given in Figs. 3.2 and 3.3 (the corresponding initial conditions are presented in the captions).

This numerical solutions show that due to the coupling with the plasma waves, the nonlinear electromagnetic waves exhibit both amplitude and phase

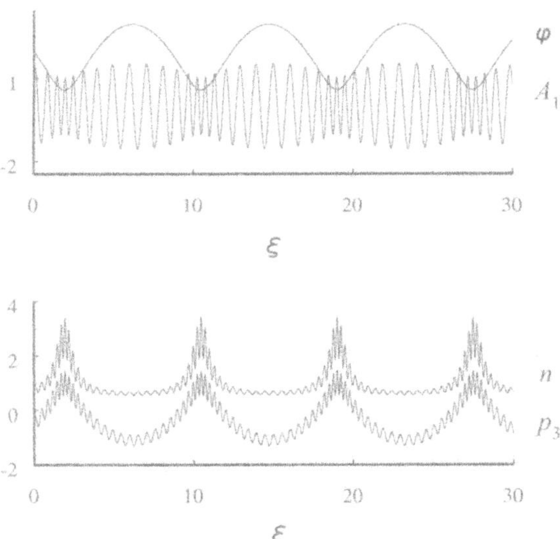

Fig. 3.1. Vector and scalar potential solutions of the Akhiezer–Polovin problem (3.8)–(3.10) and the corresponding electron longitudinal momentum and concentration calculated using (3.6)–(3.7). $A_1(0) = 1.2$, $A_{1_\xi}(0) = 5$, $\varphi(0) = 2$, $\varphi_\xi(0) = -1$, $\epsilon = 0.1$

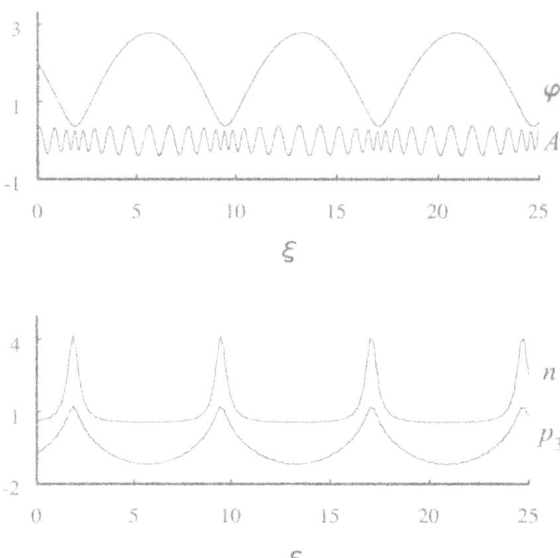

Fig. 3.2. Vector and scalar potential solutions of the Akhiezer–Polovin problem (3.8)–(3.10) and the corresponding electron longitudinal momentum and concentration calculated using (3.6)–(3.7). $A_1(0) = 0.3$, $A_{1_\xi}(0) = 1.5$, $\varphi(0) = 2$, $\varphi_\xi(0) = -1$, $\epsilon = 0.1$

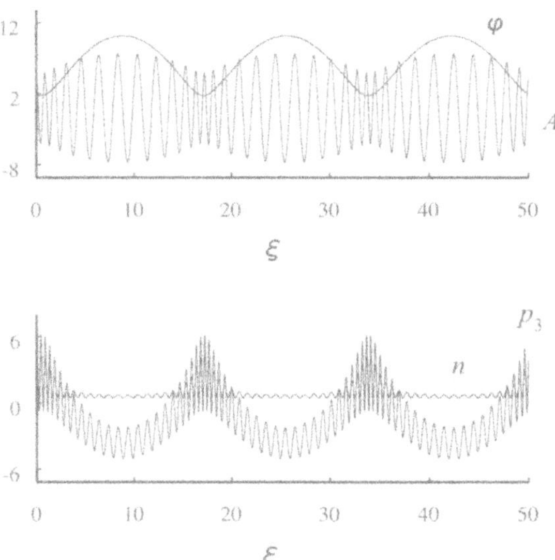

Fig. 3.3. Vector and scalar potential solutions of the Akhiezer–Polovin problem (3.8)–(3.10) and the corresponding electron longitudinal momentum and concentration calculated using (3.6)–(3.7). $A_1(0) = 5$, $A_{1_\xi}(0) = 7$, $\varphi(0) = 2$, $\varphi_\xi(0) = -1$, $\epsilon = 0.1$

self-modulation; the electromagnetic radiation is concentrated between the maxima of the electrostatic potential, and the frequency of the vector potential oscillation varies locally between the maximum and the minimum of the scalar potential.

3.2.2 Asymptotic Theory in the High-Frequency Limit

An analytical theory of nonmonochromatic high-frequency linearly polarized electromagnetic waves in plasmas was established in [43,44]. As already mentioned, generally, the phase velocity of electromagnetic waves in cold underdense plasmas (i.e., when the electromagnetic radiation frequency is much greater than the plasma frequency) is close to the speed of light, so that the parameter ϵ introduced above must be small. Under this condition, (3.8)–(3.10) comprise a singularly perturbed problem. The following ansatz was used in [43,44] to develop asymptotic solutions of the Akhiezer–Polovin problem in this case:

$$A_1 = \sum_{m=0}^{\infty} \epsilon^m \, U_m(\xi, \Theta(\xi,\epsilon)) \,, \qquad \varphi = \sum_{m=0}^{\infty} \epsilon^m \, \phi_m(\xi, \Theta(\xi,\epsilon)) \,,$$

$$\Theta_\xi = \epsilon^{-1} \, \mu(\xi) \,.$$

Here, $\mu(\xi)$ is an additional function to be defined below, and Θ plays the role of an independent variable. Substitution of these asymptotic series in the Akhiezer–Polovin problem and expansion in ϵ up to the fourth order leads to a set of partial differential equations:

$$\mu^2(\xi) U_{0_{\Theta\Theta}} + \phi_0^{-1} U_0 = 0 \,, \tag{3.11}$$

$$2\mu(\xi) U_{0_{\Theta\xi}} + \mu_\xi(\xi) U_{0_\Theta} + \mu^2(\xi) U_{1_{\Theta\Theta}} + \phi_0^{-1} U_1 - \phi_0^{-2} U_0 \phi_1 = 0 \,, \tag{3.12}$$

$$\begin{aligned} &U_{0_{\xi\xi}} + 2\mu(\xi) U_{1_{\Theta\xi}} + \mu_\xi(\xi) U_{1_\Theta} + \mu^2(\xi) U_{2_{\Theta\Theta}} + \frac{U_2}{\phi_0} - \frac{U_0 \phi_2}{\phi_0^2} \\ &+ \frac{1}{2}\left(1 - \frac{1+U_0^2}{\phi_0^2}\right) \frac{U_0}{\phi_0} - \frac{U_1 \phi_1}{\phi_0^2} + \frac{U_0 \phi_1^2}{\phi_0^3} = 0 \,, \end{aligned} \tag{3.13}$$

$$\mu^2(\xi) \phi_{0_{\Theta\Theta}} = 0 \,, \tag{3.14}$$

$$2\mu(\xi) \phi_{0_{\Theta\xi}} + \mu_\xi(\xi) \phi_{0_\Theta} + \mu^2(\xi) \phi_{1_{\Theta\Theta}} = 0 \,, \tag{3.15}$$

$$\phi_{0_{\xi\xi}} + 2\mu(\xi) \phi_{1_{\Theta\xi}} + \mu_\xi(\xi) \phi_{1_\Theta} + \mu^2(\xi) \phi_{2_{\Theta\Theta}} + \frac{1}{2}\left(1 - \frac{1+U_0^2}{\phi_0^2}\right) = 0 \,. \tag{3.16}$$

As follows from (3.14), ϕ_0 must be independent of Θ. Furthermore, combining this fact and (3.15), we find that the same is true of ϕ_1. Now, from (3.11), $U_0(\xi, \Theta) = a_0(\xi)\sin\Theta$, and $\mu(\xi) = \phi_0^{-1/2}(\xi)$. To avoid a secular dependence of U_1 on Θ, we have to assume that the right-hand side of (3.12) is equal to zero. This condition yields $\phi_1 \equiv 0$, and, therefore, $a_0(\xi) = g\,\phi_0^{1/4}$, where g is a constant. Finally, to the lowest order in ϵ,

$$U_0(\xi) = g\,\phi_0^{1/4}(\xi)\,\sin\Theta\,, \tag{3.17}$$

$$\Theta(\xi) = \frac{1}{\epsilon}\int \phi_0^{-1/2}(\xi)\,\mathrm{d}\xi\,, \tag{3.18}$$

$$\phi_{0_{\xi\xi}} = \frac{1}{2}\left(\frac{1+\frac{1}{2}g^2\phi_0^{1/2}}{\phi_0^2} - 1\right). \tag{3.19}$$

The constant g can be interpreted as the parameter of coupling of the electromagnetic field and the Langmuir plasma wake. Higher order approximations in ϵ are calculated on the basis of (3.13)–(3.16):

$$U_1(\xi, \Theta) = \frac{\phi_0^{1/4}}{2}\int \frac{g\left[g^2\,\phi_0^{1/2} - 3\,(E\phi_0 - 3\phi_0^2 + 1)\right]}{16\,\phi_0^{5/2}}\,\mathrm{d}\xi\,\cos\Theta\,,$$

$$U_2(\xi, \Theta) = \frac{g^3}{128\,\phi_0^{5/4}(\xi)}\,\sin 3\Theta + a_2(\xi)\,\cos(\Theta + \mathrm{const})\,,$$

$$\phi_2(\xi, \Theta) = \frac{g^2}{16\,\phi_0^{1/2}(\xi)}\,\cos 2\Theta\,.$$

The last term in the expression for U_2 corresponds to a small "central frequency" correction which has to be calculated with the help of higher order approximations.

To the lowest order in ϵ, the electron fluid concentration and longitudinal momentum are given by [43, 44]

$$n = 1 + \frac{1}{2}\left(\frac{1+\frac{1}{2}g^2\,\phi_0^{1/2}}{\phi_0^2} - 1\right) - \frac{g^2}{4\phi_0^{3/2}}\,\cos 2\Theta\,,$$

$$p_0 = \phi_0\,(n_0 - 1)\,. \tag{3.20}$$

Thus, the plasma electron component exhibits a slow response to the propagating electromagnetic radiation and second harmonics oscillations.

The following first integral is associated with (3.19)

$$\phi_{0_\xi}^2 + V(\phi_0) = E\,, \tag{3.21}$$

$$V(\phi_0) = V_\mathrm{p}(\phi_0) + V_\mathrm{l}(\phi_0)\,, \tag{3.22}$$

$$V_{\rm p}(\phi_0) = \phi_0 + \phi_0^{-1}\,, \tag{3.23}$$

$$V_{\rm l}(\phi_0) = g^2\,\phi_0^{-1/2}\,. \tag{3.24}$$

Two simple analogies are related to (3.17) and (3.18). Note that a formal expansion of the nonlinearity in (3.8) in ϵ^2 results in

$$\epsilon^2\,\boldsymbol{A}_{\xi\xi} + \varphi^{-1}\boldsymbol{A} = 0\,. \tag{3.25}$$

First, if we assume that φ varies slowly [this is a natural assumption since expanding (3.9) in ϵ^2 formally, one obtains an equation without singular perturbations], the above equation can be interpreted as the Einstein pendulum problem (i.e., a pendulum with a slowly varying frequency) [140]. Even though, unlike the pendulum problem in our case, this variable frequency is not an explicitly defined function of "time" but must be calculated using another nonlinear equation, the same relation between the local frequency $\Omega = \Theta_\xi = \epsilon^{-1}\phi^{-1/2}$ and amplitude $a_0 = g\phi^{1/4}$ holds true: $a_0^2\Omega = g^2/\epsilon$. As follows from the last equation, the quantity g^2/ϵ can be interpreted as the value of the adiabatic invariant for the problem embodied in (3.8)–(3.10).

Second, (3.25) is formally similar to the Schroedinger equation for a particle in a one-dimensional potential well. In the framework of this analogy, ϵ plays the role of Plank's constant, and φ^{-1} corresponds to the particle momentum squared. Again, despite the actual nonlinearity of the problem considered, (3.17) and (3.18) are readily identified as the analog of the WKB approximation.

A quantization condition is related to the approximate solutions developed above, namely,

$$J = \oint \phi^{-1/2}\,\mathrm{d}\xi = \oint \frac{\mathrm{d}\phi}{\sqrt{\phi\,[E - V(\phi)]}} = 2\pi\epsilon N\,, \tag{3.26}$$

where $N \gg 1$ is an integer. This equation can be viewed as the dispersion relation for nonlinear electromagnetic waves in cold underdense plasmas. Averaging the local electromagnetic wave frequency over the period of the "slow" plasmon, $\overline{\Omega} = J/\epsilon T = 2\pi N/T$; T is the plasmon period given by

$$T(E,g) = \oint \frac{\mathrm{d}\phi}{\sqrt{E - V(\phi)}}\,.$$

The average wavenumber for the electromagnetic wave can be defined as

$$k_0 = (T\epsilon)^{-1}\oint \phi_0^{-1/2}\,\mathrm{d}\xi\,.$$

In this notation, the dispersion relation assumes the form

$$k_0 = N\,k_{\rm p}\,, \qquad k_{\rm p} = \frac{2\pi}{T}\,.$$

The quantization condition makes it possible to calculate the intensity of the electromagnetic wave described by a solution of the Akhiezer–Polovin problem, averaged over fast electromagnetic field oscillations and over the plasmon period. To the lowest order in ϵ, the normalized electric and magnetic fields and the normalized intensity of an electromagnetic wave in a plasma are related to the scalar potential in the following way:

$$\boldsymbol{E} = -\boldsymbol{A}_t - \nabla\varphi = \boldsymbol{e}_1 \frac{g}{\epsilon\varphi^{1/4}} \cos\Theta + \dots ,$$

$$\boldsymbol{B} = [\nabla, \boldsymbol{A}] = \boldsymbol{e}_2 \frac{g}{\epsilon\varphi^{1/4}} \cos\Theta + \dots ,$$

$$\boldsymbol{I} = [\boldsymbol{E}, \boldsymbol{B}] = \mathbf{e}_3 I , \qquad I = \frac{g^2}{\epsilon\varphi^{1/2}} \cos^2\Theta + \dots .$$

The average of this intensity over fast oscillations of the electromagnetic field makes

$$I_{\text{slow}} = \frac{g^2}{2\,\epsilon\varphi^{1/2}} + \dots .$$

Next, averaging this slowly varying quantity over the plasmon period T and using the quantization condition (3.26),

$$\overline{I_{\text{slow}}} = T^{-1} \oint I_{\text{slow}}\, \mathrm{d}\xi = \frac{g^2\, k_0}{2\,\epsilon} . \tag{3.27}$$

The normalized averaged intensities of the electromagnetic waves depicted in Figs. 3.1–3.3 are 51.5, 3.8, and 390, respectively.

It can be seen easily that the simplest solution of (3.19) is given by a constant $\phi \equiv \phi_0$. The corresponding electromagnetic wave–plasmon coupling parameter makes

$$g = \frac{\sqrt{2(\phi_0^2 - 1)}}{\phi_0^{1/4}}$$

(naturally, $\phi_0^2 > 1$ must be required). In this particular case, to the lowest order in ϵ, the electromagnetic field is monochromatic:

$$U_0 = \sqrt{2(\phi_0^2 - 1)}\, \sin\left(\frac{\xi}{\epsilon\sqrt{\phi_0}}\right) , \tag{3.28}$$

and $\phi_0^2 = 1 + \overline{U_0^2}$. Figure 3.4 shows a nearly monochromatic numerical solution of (3.8)–(3.10) obtained by arranging for the combination of the problem parameters in accord with the above recipe. The linear instability of a solution of this type is studied in [66].

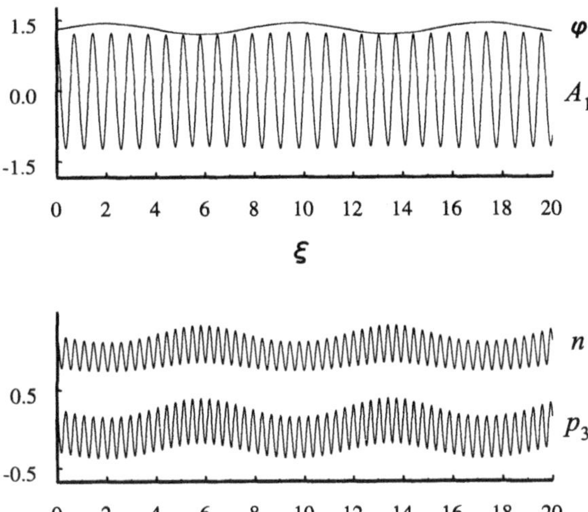

Fig. 3.4. Vector and scalar potential solutions of the Akhiezer–Polovin problem (3.8)–(3.10) and the corresponding electron longitudinal momentum and concentration calculated using (3.6)–(3.7). $A_1(0) = 1.2$, $A_{1_\xi}(0) = 0$, $\varphi(0) = \sqrt{1 + A_1^2(0)}$, $\varphi_\xi(0) = 0$, $\epsilon = 0.1$. The first plot represents a nearly monochromatic linearly polarized plane electromagnetic wave in a plasma

3.2.3 Quasi-Relativistic Limit

Under certain conditions, the propagation of intense electromagnetic fields in cold underdense plasmas can be described by a quasi-relativistic approximation. This is the case when the electromagnetic radiation intensity is low enough so that relativistic corrections of the masses of free electrons driven by the electromagnetic fields are adequately accounted for by terms of third order in the ratio of the electron momentum and the rest momentum, whereas higher order terms can be neglected (here attention should be paid to the fact that the phase velocity in the Akhiezer–Polovin problem is greater than the speed of light). The most rigorous treatment of this issue can be found in [141], where a small parameter is introduced to characterize the deviation of the field and plasma parameters from their unperturbed values and a system of new nonlinear equations is derived by expansions in this parameter (also, see [142]). The quasi-relativistic field–plasma equations are used in [59] to investigate the instability of circularly polarized electromagnetic waves in slab geometry.

Below, we consider a partial case of the system of equations proposed in [141], in which the quasi-relativistic approximation of the Akhiezer–Polovin problem is embodied. The equations considered below can be obtained by assuming that the time derivatives in the set of equations presented in [141] and [142] are equal to zero. Alternatively, one can derive the quasi-relativistic equations directly from (3.8)–(3.10) by expanding the nonlinear terms in ϵ^2, presenting the scalar potential in the form

$$\varphi = 1 + f , \tag{3.29}$$

assuming that $f \sim A_1^2$ (here we continue to investigate linearly polarized electromagnetic waves, so that $A_2 \equiv 0$) and neglecting the corrections of order A_1^4 and higher. The result is as follows:

$$\epsilon^2 A_{1_{\xi\xi}} + (1 - f)A_1 = 0 , \tag{3.30}$$

$$f_{\xi\xi} + f = \frac{1}{2} A_1^2 . \tag{3.31}$$

One of the potentially useful techniques for treating these equations is the stationary pump approximation. For a powerful electromagnetic field with a wide amplitude spacial distribution, one can assume that the shape of this distribution is preserved during laser pulse propagation. Then, the second of the above equations is linear with respect to the scalar potential, and its solution is [147]

$$f(\xi) = \int_{-\infty}^{\infty} \sin |\xi - \xi'| \frac{A_1^2(\xi')}{2} \, d\xi'$$

(provided that the integral converges).

Approximate solutions of problem (3.30)–(3.31) are developed in the same way as for the fully relativistic case explored in the previous section; the result is

$$A_1 = g\,(1 - f)^{-1/4} \sin \left[\epsilon^{-1} \int (1 - f)^{1/2} \, d\xi \right] + O(\epsilon^2) , \tag{3.32}$$

$$f_{\xi\xi} + f = \frac{1}{4} g^2 (1 - f)^{1/2} .$$

The last equation is integrated easily:

$$f_\xi^2 + V_{qr}(f) = E , \tag{3.33}$$

$$V_{qr}(f) = f^2 + g^2 (1 - f)^{1/2} . \tag{3.34}$$

Naturally, the above approximate expressions can be derived from the corresponding fully relativistic ones. Substituting (3.29) in (3.17)–(3.18) and performing an elementary transformation, to the lowest order in the small parameter ϵ,

$$A_1 = g \left(\frac{1 - f^2}{1 - f} \right)^{1/4} \sin \left[\epsilon^{-1} \int \left(\frac{1 - f^2}{1 - f} \right)^{-1/2} d\xi \right] .$$

In accord with the quasi-relativistic approximation, f^2 is negligible, and we arrive at (3.32). Further, since $V_p(\varphi) = V_p(1 + f) \approx f^2$ and, obviously, $V_l(\varphi) = V_l(1 + f) = g^2 \sqrt{(1 - f)(1 - f^2)}$, once again neglecting f^2 we obtain (3.33) and (3.34) as a consequence of the fully relativistic conservation law (3.21)–(3.24).

3.3 Circularly Polarized Plane Electromagnetic Waves

Asymptotic solutions of the Akhiezer–Polovin problem (3.8)–(3.10) for circularly polarized electromagnetic waves in plasmas are derived in the same way as those for linear polarization; the lowest order result is

$$A_1(\xi) = g\,\phi_0^{1/4}(\xi)\,\sin\Theta + O(\epsilon)\,,$$

$$A_2(\xi) = g\,\phi_0^{1/4}(\xi)\,\cos\Theta + O(\epsilon)\,,$$

and the equation for φ and its first integral differ from those given by (3.19), (3.21)–(3.24) in the following way: there is g^2 instead of $g^2/2$ and $2g^2$ instead of g^2 in them, respectively. In this case, there is no second-harmonics term in the plasma wake equation

$$n_0 - 1 = \frac{1}{2}\left(\frac{1 + g^2\,\phi_0^{1/2}}{\phi_0^2} - 1\right),$$

[the electron fluid momentum is related to the above quantity by (3.20) just as in linear polarization].

Also, there exist monochromatic circularly polarized waves given by

$$A_1(\xi) = \sqrt{\phi_0^2 - 1}\,\sin\frac{\xi}{\epsilon\sqrt{\phi_0}}\,, \tag{3.35}$$

$$A_2(\xi) = \sqrt{\phi_0^2 - 1}\,\cos\frac{\xi}{\epsilon\sqrt{\phi_0}} \tag{3.36}$$

(it is assumed that $\phi_0^2 > 1$). In more conventional notations,

$$a_0 = \sqrt{\phi_0^2 - 1}\,, \qquad k_0 = \frac{\xi}{\epsilon\sqrt{\phi_0}}\,, \qquad \gamma_0^{-1} = \phi_0^{-1}\,, \tag{3.37}$$

the latter solution becomes identical to that examined in the previous chapter [see (2.65)–(2.66)].

Note that this particular solution is exact. A number of works was dedicated to its instability (for example, see [44, 62, 63]), which will be considered later as well.

4. Instabilities of Circularly Polarized Plane Electromagnetic Waves in Plasmas

Various types of relativistically intense electromagnetic waves in cold underdense plasmas, which we explored in the previous chapter, can be unstable to small perturbations. Investigation of these instabilities makes it possible to describe the scattering of powerful laser radiation in plasmas, harmonics generation, and a whole ensemble of other phenomena. Since extremely high intensities are considered, the very first requirement, which, for example, is met by employing solutions of the Akhiezer–Polovin problem, is to use arbitrary amplitude solutions of the Maxwell and plasma dynamics equations as the ground state for the perturbative analysis.

Two approaches to the development of such a theory are possible. First, one can consider the evolution of small perturbations in the field–plasma system against the background given by the solutions of the Akhiezer–Polovin problem. Second, simplified model equations can be derived, and the perturbation theory can be developed in their framework. Each of the methods has its advantages and drawbacks. A general instability theory which can be established using the former approach describes an extensive range of phenomena such as electromagnetic radiation generation at harmonics of the propagating wave, the continuum of scattered radiation, polarization violation, etc., and, importantly, the interplay of all of these effects. However, the latter model approach should make it possible to present a theory accounting for laser pulses of finite size and duration, which obviously cannot be done otherwise.

A general study of the instability of the solutions to the Akhiezer–Polovin problem describing circularly polarized relativistically intense electromagnetic radiation is performed in this chapter. The next one is dedicated to the results of a similar inquiry involving linearly polarized relativistically intense electromagnetic waves in cold underdense plasmas. The perturbation evolution against the background defined by such solutions obeys linearized equations with oscillating coefficients, so that a special technique is needed to calculate the corresponding growth rates. In the Fourier space, these equations become an infinite-dimension set of coupled ordinary differential equations. When approximating this set by a finite-dimension one to calculate the eigenvalues numerically it is necessary to verify that this finite dimension is sufficiently high. This is accomplished by repeating calculations with yet bigger matrices and making sure that this does not lead to substantial changes in the outcome of the computational procedure.

Results of studies of the relativistically intense laser pulse instabilities in a model framework are summarized in Chap. 6.

4.1 Equations of Circularly Polarized Wave Instability in Plasmas

We use a solution of the Akhiezer–Polovin problem (3.8)–(3.10) as the ground state for the linear analysis of the instability to be performed in the framework of (2.56)–(2.61). Below, the functions corresponding to the Akhiezer–Polovin problem solutions are denoted by subscript "0" and the perturbations are denoted by $\delta\boldsymbol{A}$, $\delta\varphi$, δn, $\delta\boldsymbol{p}$, so that

$$
\begin{aligned}
\boldsymbol{A} &= \boldsymbol{A}_0 + \delta\boldsymbol{A}\,, \qquad & \varphi &= \varphi_0 + \delta\varphi\,, \\
n &= n_0 + \delta n\,, \qquad & \boldsymbol{P} &= \boldsymbol{P}_0 + \delta\boldsymbol{P}\,.
\end{aligned}
$$

The linearized equations governing the perturbations are

$$
\begin{aligned}
(\triangle - \partial_t^2)\,\delta\boldsymbol{A} = \nabla\delta\varphi_t &+ \frac{n_0}{\gamma_0}(\delta\boldsymbol{A}+\delta\boldsymbol{P}) + \frac{1}{\gamma_0}(\boldsymbol{A}_0+\boldsymbol{P}_0)\,\delta n \\
&- \frac{n_0}{\gamma_0^3}(\boldsymbol{A}_0+\boldsymbol{P}_0)\,(\boldsymbol{A}_0+\boldsymbol{P}_0, \delta\boldsymbol{A}+\delta\boldsymbol{P})\,,
\end{aligned}
\tag{4.1}
$$

$$
(\nabla, \delta\boldsymbol{A}) = 0,\ \triangle\,\delta\varphi = \delta n\,, \tag{4.2}
$$

$$
\delta\boldsymbol{P}_t = \nabla\left(\delta\varphi - \gamma_0^{-1}(\boldsymbol{A}_0+\boldsymbol{P}_0, \delta\boldsymbol{A}+\delta\boldsymbol{P})\right)\,, \tag{4.3}
$$

$$
\begin{aligned}
\delta n_t + \Bigg[\nabla, &\frac{n_0}{\gamma_0}(\delta\boldsymbol{A}+\delta\boldsymbol{P}) + \frac{1}{\gamma_0}(\boldsymbol{A}_0+\boldsymbol{P}_0)\,\delta n \\
&- \frac{n_0}{\gamma_0^3}(\boldsymbol{A}_0+\boldsymbol{P}_0)\,(\boldsymbol{A}_0+\boldsymbol{P}_0, \delta\boldsymbol{A}+\delta\boldsymbol{P})\Bigg] = 0\,.
\end{aligned}
\tag{4.4}
$$

In this chapter, we consider the instability of circularly polarized electromagnetic waves given by (3.35)–(3.36) or, which is the same, by (2.65)–(2.68).

The co-moving variable can be introduced; the result is a set of linear partial differential equations whose coefficients are periodic in ξ. Since the propagation of radiation in an unbounded plasma is considered, we Fourier-transform this set of equations in $\boldsymbol{x}_\perp$ and ξ:

$$
\begin{aligned}
(\delta\boldsymbol{A}, \delta\varphi, \delta n, \delta\boldsymbol{p})^T = (2\pi)^{-3/2} &\int \exp(\mathrm{i}\boldsymbol{k}, (x_1, x_2, \xi)) \\
&\times (\delta\boldsymbol{A}, \delta\varphi, \delta n, \delta\boldsymbol{p})^T(\boldsymbol{k}, t)\,\mathrm{d}^3 k\,.
\end{aligned}
\tag{4.5}
$$

The equations for the Fourier transforms are

$$\hat{D}\,\delta A_1 + g_1 \frac{ik_1\chi^2}{\boldsymbol{k}_\perp^2+\chi^2}\,\delta\psi - g_1 \frac{k_1k_2}{\boldsymbol{k}_\perp^2+\chi^2}\,\delta A_2 + g_1 \frac{k_2^2+\chi^2}{\boldsymbol{k}_\perp^2+\chi^2}\,\delta A_1$$
$$+g_2\left(F^-_{1,1_{\chi-k_0}} + F^+_{1,1_{\chi+k_0}}\right) + \frac{g_1}{2}\left(F^-_{1,2_{\chi-2k_0}} + F^+_{1,2_{\chi+2k_0}}\right) = 0\,, \tag{4.6}$$

$$\hat{D}\,\delta A_2 + g_1 \frac{ik_2\chi^2}{\boldsymbol{k}_\perp^2+\chi^2}\,\delta\psi - g_1 \frac{k_1k_2}{\boldsymbol{k}_\perp^2+\chi^2}\,\delta A_1 + g_1 \frac{k_1^2+\chi^2}{\boldsymbol{k}_\perp^2+\chi^2}\,\delta A_2$$
$$+g_2\left(F^-_{2,1_{\chi-k_0}} + F^+_{2,1_{\chi+k_0}}\right) + \frac{g_1}{2}\left(F^-_{2,2_{\chi-2k_0}} + F^+_{2,2_{\chi+2k_0}}\right) = 0\,, \tag{4.7}$$

$$-(\boldsymbol{k}_\perp^2+\chi^2)\varphi = n\,, \tag{4.8}$$

$$\hat{D}_t\,\delta\psi - \delta\varphi + g_2\left[(\Pi_1+\mathrm{i}\Pi_2)_{\chi-k_0} + (\Pi_1-\mathrm{i}\Pi_2)_{\chi+k_0}\right] = 0\,, \tag{4.9}$$

$$\begin{aligned}&\hat{D}_t\,\delta n - \left[\gamma_0^{-1}\chi^2 + (\gamma_0^{-1} - g_1\boldsymbol{k}_\perp^2)\right]\delta\psi - \mathrm{i}g_1(k_1\delta A_1 + k_2\delta A_2)\\ &+g_2\left[(\mathrm{i}k_1-k_2)\delta n_{\chi-k_0} + (\mathrm{i}k_1+k_2)\delta n_{\chi+k_0}\right]\\ &-\frac{g_1}{2}\left[(\mathrm{i}k_1-k_2)\Pi_1 - (\mathrm{i}k_2+k_1)\Pi_2\right]_{\chi-2k_0}\\ &-\frac{g_1}{2}\left[(\mathrm{i}k_1+k_2)\Pi_1 - (\mathrm{i}k_2-k_1)\Pi_2\right]_{\chi+2k_0} = 0\,.\end{aligned} \tag{4.10}$$

The following notations are used in the above equations:

$$g_1 = \frac{a_0^2}{2\gamma_0^3}\,, \qquad g_2 = \frac{a_0}{2\gamma_0}\,,$$

$$\Pi_{1,2} = \delta A_{1,2} + \mathrm{i}k_{1,2}\,\delta\psi\,,$$

$$\hat{D}_t = \partial_t - \mathrm{i}q\chi\,, \qquad \hat{D} = -\partial_t^2 + 2\mathrm{i}q\chi\,\partial_t + \left(\frac{\chi^2}{\gamma_0k_0^2} - \boldsymbol{k}_\perp^2\right) - \gamma_0^{-1}\,,$$

$$F^\pm_{1,1} = \left(k_1\frac{k_1\mp\mathrm{i}k_2}{\boldsymbol{k}_\perp^2+\chi^2} - 1\right)n\,,$$

$$F^\pm_{1,2} = \left(1 - k_1\frac{k_1\mp\mathrm{i}k_2}{\boldsymbol{k}_\perp^2+\chi^2}\right)(\Pi_1\mp\mathrm{i}\Pi_2)\,,$$

$$F^\pm_{2,1} = -\mathrm{i}\left(k_2\frac{k_1\mp\mathrm{i}k_2}{\boldsymbol{k}_\perp^2+\chi^2} \mp 1\right)n\,,$$

$$F^\pm_{2,2} = \mathrm{i}\left(k_2\frac{k_1\mp\mathrm{i}k_2}{\boldsymbol{k}_\perp^2+\chi^2} \mp 1\right)(\Pi_1\pm\mathrm{i}\Pi_2)\,.$$

Indexes $\chi \pm nk_0$ denote the shifted arguments of the corresponding functions. Rewriting the above equations for a sequence of arguments given by $\chi \pm lk_0$ where l is an integer, we establish an infinite set of coupled linear ordinary differential equations for the amplitudes of laser radiation perturbation harmonics in the Fourier-transform space. This set can be cast in the form

$$Y_t = BY\,, \tag{4.11}$$

where Y is an infinite column and B is an infinite-dimension, 30-diagonal matrix. Below, the eigenvalues of matrix B will be calculated numerically with the help of the techniques presented in [143]. The growth rate G of the relativistically intense electromagnetic radiation instability is defined as the maximal real part of the eigenvalue of matrix B. Naturally, the infinite-dimension matrix B must be approximated by a finite-dimension matrix to perform computations. This approximation introduces a distortion into the calculated eigenvalue, so that a series of calculations with a range of finite dimensions is needed to find out whether the truncated matrix used is large enough.

4.2 Slab Geometry Instability Equations

As the first step, let us examine the case of large transverse aperture perturbations, when we can set $k_1 = k_2 = 0$; than (4.6)–(4.10) become

$$\begin{aligned}&\hat{D}_1\,\delta A_1 + g_1\,\delta A_1 - g_2(\delta n_{\chi-k_0} + \delta n_{\chi+k_0})\\&+\frac{g_1}{2}\left[(\delta A_1 + \mathrm{i}\delta A_2)_{\chi-2k_0} + (\delta A_1 - \mathrm{i}\delta A_2)_{\chi-2k_0}\right] = 0\,,\end{aligned} \tag{4.12}$$

$$\begin{aligned}&\hat{D}_1\,\delta A_2 + g_1\,\delta A_2 - \mathrm{i}g_2(\delta n_{\chi-k_0} - \delta n_{\chi+k_0})\\&+\frac{g_1}{2}\left[(\mathrm{i}\delta A_1 - \delta A_2)_{\chi-2k_0} - (\mathrm{i}\delta A_1 - \delta A_2)_{\chi-2k_0}\right] = 0\,,\end{aligned} \tag{4.13}$$

$$\hat{D}_t\delta\psi + \chi^{-2}\delta n + g_2\left[(\delta A_1 + \mathrm{i}\delta A_2)_{\chi-k_0} + (\delta A_1 - \mathrm{i}\delta A_2)_{\chi+k_0}\right] = 0\,, \tag{4.14}$$

$$\hat{D}_t\,\delta n = \gamma_0^{-1}\chi^2\,\delta\psi\,, \tag{4.15}$$

where

$$\hat{D}_1 = -\partial_t^2 + 2\mathrm{i}q\chi\partial_t + \frac{\chi^2}{\gamma_0 k_0^2} - \gamma_0^{-1}\,.$$

Like the general case, the above equations can be presented in the form given by (4.11). The structure of the corresponding matrix B is similar to that of the three-dimensional case.

4.2.1 Conserved Circular Polarization Approximation

In contrast to the equations used in [61, 121], (4.6)–(4.10) and their slab geometry analogs (4.12)–(4.15) describe electromagnetic field perturbations of arbitrary polarization. In the slab geometry case, a partial solution of (4.6)–(4.10) exists, which describes a circularly polarized vector potential, but the assumption of circular polarization of perturbations [61, 121] results in a substantial loss of generality. Let us demonstrate that under certain conditions the slab geometry equations (4.12)–(4.15) are reduced to those for circularly polarized electromagnetic field perturbations.

In slab geometry for circularly polarized perturbations of the electromagnetic radiation,

$$\boldsymbol{A} = \frac{1}{2}(\boldsymbol{e}_1 + i\boldsymbol{e}_2)\, a(\xi, t) \exp(ik_0\xi) + c.c.\,,$$

so that in the Fourier-transform space

$$\delta A_1 = \frac{1}{2}(\delta a_{\chi-k_0} + \delta b_{\chi+k_0}),\ \delta A_1 = \frac{i}{2}(\delta a_{\chi-k_0} - \delta b_{\chi+k_0})\,, \tag{4.16}$$

$$\delta b(\chi, t) = \delta a^*(-\chi, t)\,,$$

where a is the slow amplitude of the electromagnetic field. Substituting these relations in (4.12) and (4.13), we arrive at the following equation for this function:

$$\begin{aligned} &-\delta a_{tt} + 2iq\chi\,\delta a_t + \frac{\chi^2}{\gamma_0 k_0^2}\,\delta a + 2i\left(\omega_0\,\delta a_t - \frac{i\chi}{\gamma_0 k_0}\,\delta a\right) \\ &= 2g_2\,\delta n - g_1(\delta a + \delta b)\,. \end{aligned} \tag{4.17}$$

Differentiating (4.15) in time, using (4.14), and expressing a and b from (4.16), we derive the linearized equation for the plasma wake generated by the propagating radiation:

$$(\hat{D}_t^2 + \gamma_0^{-1})\,\delta n = -\frac{a_0\chi^2}{2\gamma_0^2}\,(\delta a + \delta b)\,. \tag{4.18}$$

Equations (4.17) and (4.18) are the Fourier transforms of the equations used in [61] to study the instability of laser radiation in plasma in the framework of the conserved circular polarization assumption.

4.2.2 Instability Growth Rates in Slab Geometry

Consider the growth rate for problem (4.12)–(4.15). The corresponding computational results are depicted in Figs. 4.1–4.4. A range of dimensions of the matrix approximating the infinite matrix B was considered according to $l = 6 + 12j$, $j = 0, 1, 2, \ldots 17$. Higher values of j correspond to higher numbers of harmonics $2j + 1$ included in the simulation, as seen in Figs. 4.1–4.3. The pump laser radiation frequency corresponds to $j = 1$. The scattered radiation wave vector is denoted by χ. Propagation in the positive direction of the $\boldsymbol{e}_3$ axis corresponds to $\chi > 0$, and in the negative direction to $\chi < 0$.

Perturbation growth rates as functions of χ are shown in Figs. 4.3 and 4.4. The scattered radiation comprises a set of harmonics, and each of them is a doublet consisting of a Stokes and an anti-Stokes component. The local maxima in Figs. 4.3 and 4.4 are found at $\chi = \pm jk \pm k_{\mathrm{p}}$. There are also small

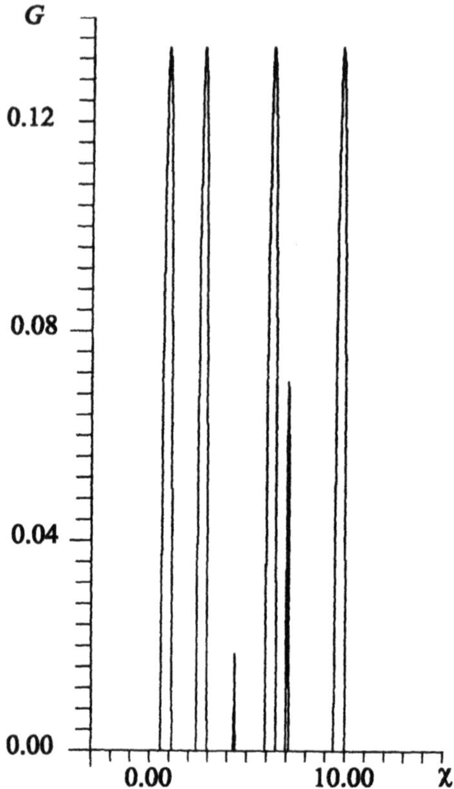

Fig. 4.1. Slab geometry. The influence of the number of harmonics included in the simulation on the growth rate. Number of harmonics: $2j+1=5$. Parameter values are $a_0 = 0.1$, $\epsilon^2 = 7.43 \times 10^{-2}$. Dependencies of the growth rate on the wave vector for $\chi > 0$ are depicted. The plot is symmetrical in χ [44]

extremums near $\chi = \pm jk$ corresponding to the fluid dynamics analog of Compton scattering.

Computations show that for different, sufficiently high values of the truncated matrix B dimension, which we denote by m, the changes in the growth rates become small in all of the harmonics except for a few highest ones (see Figs. 4.1–4.3). In other words, there is a boundary effect in simulations with a finite-dimension matrix B. For example, the $j = 0, 1, ..14$ harmonics are described adequately for $m = 210$, whereas the $j = 15, 16, 17$ harmonics are distorted by the boundary effect.

Figure 4.4 shows the simulation results corresponding to an electron concentration a factor of 2 lower than that in the case depicted in Fig. 4.3. Consequently, the plasma frequency ω_{p} and the plasma wave vector k_{p} are

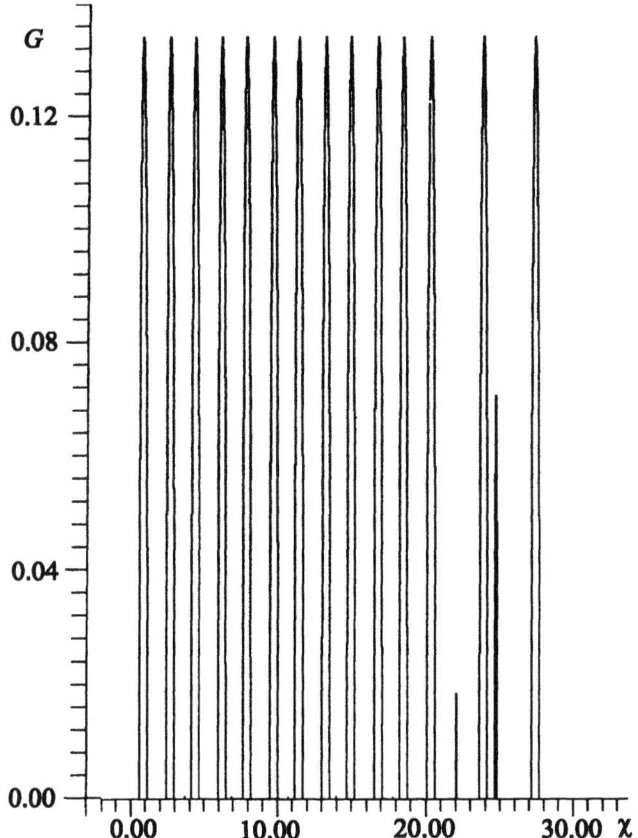

Fig. 4.2. Slab geometry. The influence of the number of harmonics included in the simulation on the growth rate. Number of harmonics: 15. Parameter values are $a_0 = 0.1$, $\epsilon^2 = 7.43 \times 10^{-2}$. Dependencies of the growth rate on the wave vector for $\chi > 0$ are depicted. The plot is symmetrical in χ [44]

a factor of $\sqrt{2}$ lower, and the Raman scattering components are located a factor of $\sqrt{2}$ closer in the case depicted in Fig. 4.4 than in the case depicted in Fig. 4.3.

As follows from Fig. 4.4, Raman scattering components are broader for higher pump electromagnetic wave intensities. Pairs of Raman scattering components merge when $a_0 \cong 1$. Therefore harmonics components are distinguishable in the relativistic case $a_0 > 1$, but the Raman scattering components can be indistinguishable.

Importantly, the understanding of the scattering bandwidths' dependencies on the propagating laser radiation intensity can be used for propagation process experimental diagnostics. It is interesting to compare the above sim-

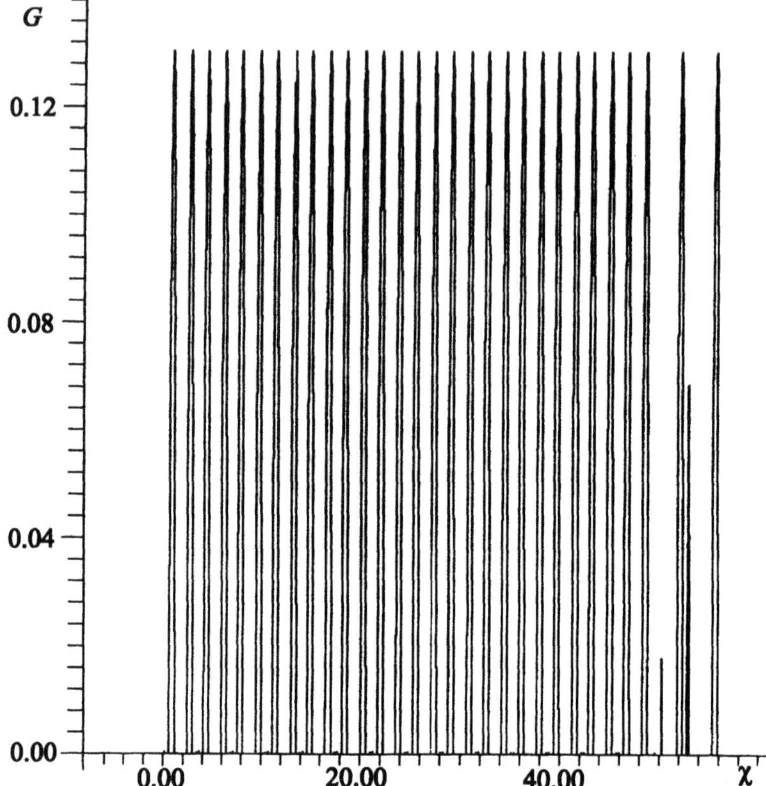

Fig. 4.3. Slab geometry. The influence of the number of harmonics included in the simulation on the growth rate. Number of harmonics: 35. Parameter values are $a_0 = 0.1$, $\epsilon^2 = 7.43 \times 10^{-2}$. Dependencies of the growth rate on the wave vector for $\chi > 0$ are depicted. The plot is symmetrical in χ [44]

ulation results against those of [61], where circular polarization of scattered radiation was assumed in the slab geometry case. As we have seen, this assumption leads to a failure to detect the infinite set of harmonics. Then, the dimension of the problem matrix is 6×6. But for an arbitrary polarization all of the harmonics are excited and interact in the slab geometry case. Mathematically, this amounts to the necessity to solve an infinite set of coupled linear differential equations. Note that the simulations performed with the matrix B of a low 8×8 dimension yield results similar to those of [61].

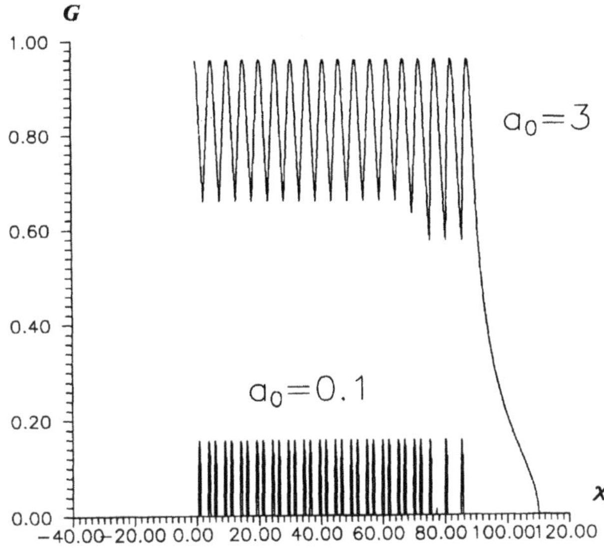

Fig. 4.4. Slab geometry. The dependence of the instability growth rate on the incident wave amplitude for χ. The plot is symmetrical in χ. Incident wave amplitudes: $a_0 = 0.1$ and $a_0 = 3$. $\epsilon^2 = 3.72 \times 10^{-2}$ [44]

4.3 3-D Instability Growth Rates

Consider the three-dimensional problem (4.6)–(4.10). Below, the growth rate of this problem is calculated as the function of the three wave vector $\boldsymbol{k} = (k_1, k_2, \chi)$ components. However, only dependencies on two variables can be plotted. For example, we consider the following growth rate distributions: $G(k_1, 0, \chi)$, $G(0, k_2, \chi)$, $G(k_1, k_2, 0)$, and $G(|\boldsymbol{k}| \cos\alpha, |\boldsymbol{k}| \sin\alpha, \chi)$, where α is an angle and $|\boldsymbol{k}| = \sqrt{k_1^2 + k_2^2 + \chi^2}$. Computations show that the growth rate is quasi-periodic in χ and that its transverse distributions are not axially symmetrical. The latter fact follows from the axial asymmetry of the linearized equations on the pump wave period. Figures 4.5–4.7 show the growth rate of the instability of a plane circularly polarized monochromatic wave for $a_0 = 0.1$ and $\epsilon = 7.4362 \times 10^{-2}$. The dependence of the growth rate on k_2 and χ for $k_1 = 0$ is depicted in Fig. 4.5. Figure 4.6 shows the same growth rate dependence on k_2 and χ for $(k_2/k_1) = 10^{-3}$. The $k_2 = 0$ section is illustrated by Fig. 4.7.

Growth rates for the instability of circularly polarized electromagnetic waves of higher amplitudes in plasmas are presented in Figs. 4.8 and 4.9 (the latter figure illustrates the instability behavior for relatively large values of k_1).

The form of the oscillating coefficients of the linearized equations (4.6)–(4.10) depends on the reference frame used. In our case, the choice of the

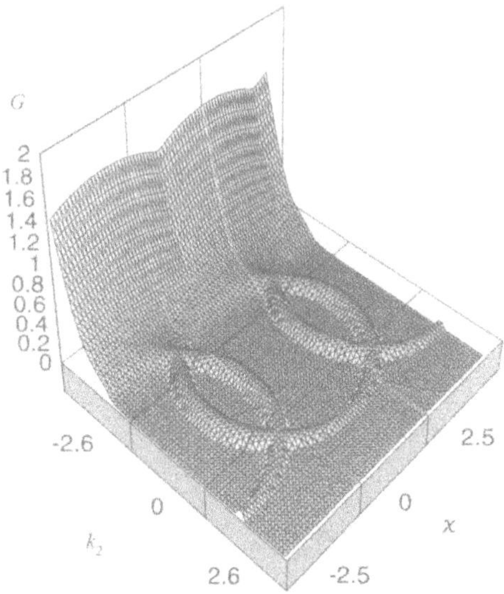

Fig. 4.5. The dependence of the instability growth rate on k_2 and χ for $k_1 = 0$. The simulation parameters are $a_0 = 0.1$, $\epsilon^2 = 7.4362 \times 10^{-2}$ [44]

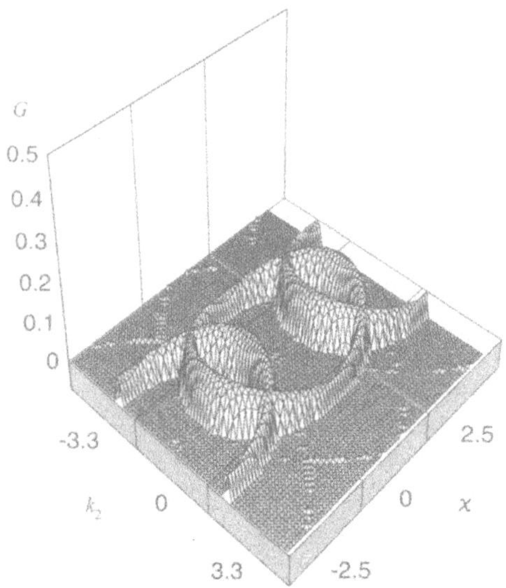

Fig. 4.6. The dependence of the instability growth rate on k_2 and χ for $(k_2/k_1) = 10^{-3}$. The simulation parameters are: $a_0 = 0.1$, $\epsilon^2 = 7.4362 \times 10^{-2}$ [44]

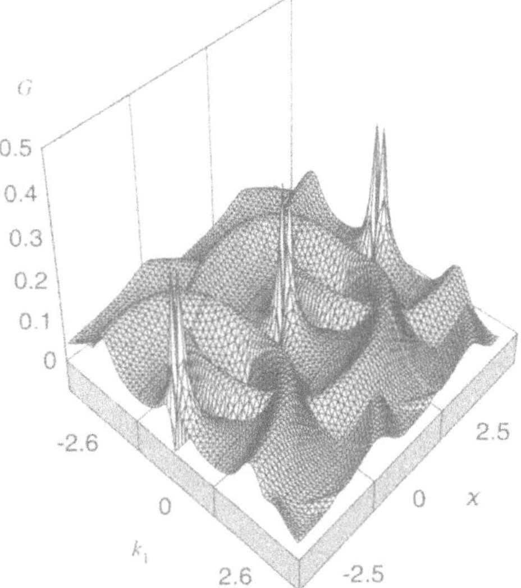

Fig. 4.7. The dependence of the instability growth rate on k_1 and χ for $k_2 = 0$. The simulation parameters are $a_0 = 0.1$, $\epsilon^2 = 7.4362 \times 10^{-2}$ [44]

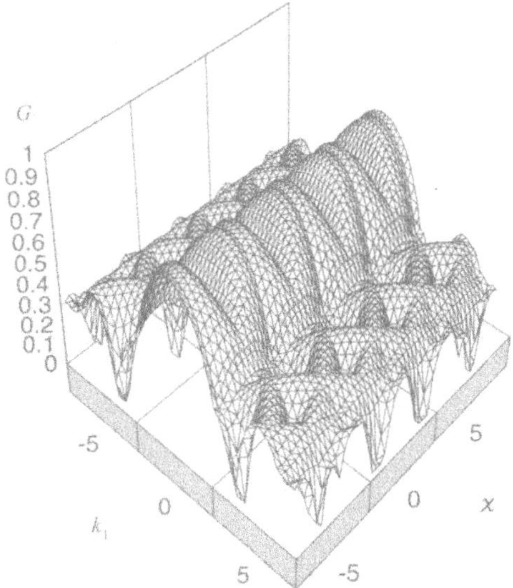

Fig. 4.8. The dependence of the instability growth rate on k_1 and χ for $k_2 = 0$. The simulation parameters are $a_0 = 1$, $\epsilon^2 = 7.4362 \times 10^{-2}$ [44]

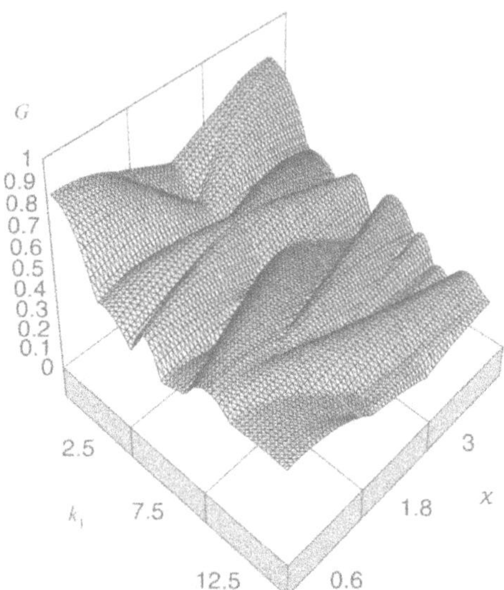

Fig. 4.9. The dependence of the instability growth rate on k_1 and χ for $k_2 = 0$. The simulation parameters are $a_0 = 3$, $\epsilon^2 = 7.4362 \times 10^{-2}$ [44]

ground state for the perturbation procedure is such that the vector $\boldsymbol{A}_0$ happens to be directed along the $\hat{\boldsymbol{e}}_1$ axis at $x_3 = 0$ for $t = 0$. But the choice of zero time within the pump wave period should be considered random. Consequently, the growth rate must be averaged over the initial time within the wave period, which is equivalent to averaging the growth rate over the azimuthal angle in the wave vector space. So the unaveraged results are intermediate and do not correspond to observable values, whereas the averaged values relate to physically observable quantities.

Figure 4.10 shows the averaged growth rate as function of $|\boldsymbol{k}|$ and χ for $a_0 = 0.1$ and $\epsilon^2 = 7.4362 \times 10^{-2}$. It is quasi-periodic in χ. Obviously, the basic growth rate features are

- a set of interconnected rings,
- repetitive maxima located near the $\boldsymbol{e}_3$ axis, and
- an increase in the growth rate for $|\boldsymbol{k}| \to \infty$.

The following details of the scattering of circularly polarized laser pulses in plasmas are illustrated in Fig. 4.10. Harmonics with wave vectors making $m\boldsymbol{k}_0 - \delta\boldsymbol{k}$, $m = 0, \pm 1, \pm 2, \ldots$ are excited in the plasma into which the pump wave propagates ($\delta\boldsymbol{k}$ is the wave vector shift due to the electron response, $|\delta\boldsymbol{k}| \ll \boldsymbol{k}_0$). Energy and momentum are conserved by the initial set of equations (see Chap. 2), and since no additional assumptions are introduced in their subsequent treatment (except for the linearization), the theory presented describes the electron response adequately. Due to the decay instability, each of the harmonics gives rise to an electromagnetic wave (the Stokes

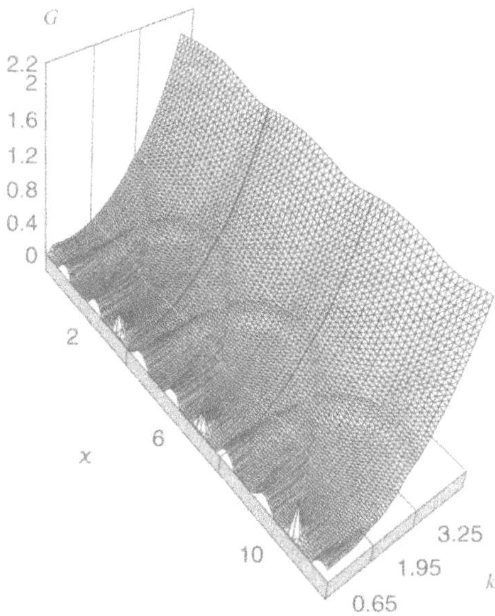

Fig. 4.10. The dependence of the two-dimensional growth rate distribution averaged over the azimuthal angle on $k_\perp$ and χ. The parameter values are $a_0 = 0.1$, $\epsilon^2 = 7.4362 \times 10^{-2}$ [44]

component of Raman scattering) and a plasma wave:

$$m\boldsymbol{k}_0 - \delta\boldsymbol{k} \to \boldsymbol{k}'_m + \boldsymbol{k}_\mathrm{p}\,, \qquad m\omega_0 - \delta\omega \to (m\omega_0 - \delta\omega - \omega_\mathrm{p}) + \omega_\mathrm{p}\,.$$

Since the cold plasma oscillation wave vector is arbitrary, the same is true of the $\boldsymbol{k}'_m$ direction in space. Thus, the spatial distribution of the growth rate is similar to a circle whose radius is $|\boldsymbol{k}'_m|$. Furthermore, wave interactions result in scattered waves with $\boldsymbol{k}'' = n\boldsymbol{k}_0 + \boldsymbol{k}'_m$, where n and m are arbitrary integers. The presence of the $\boldsymbol{k}'' = n\boldsymbol{k}_0 + \boldsymbol{k}_1$ ring structures can be seen in Fig. 4.10. The ring structures corresponding to large values of m are averaged out, but they are found in unaveraged growth rate distributions. There are no anti-Stokes components in the medium for which the ground state involves no plasma wave, where the frequency is ω_p (this is the particular feature of the ground state solution used), and computations corroborate this fact. But wave interactions between the harmonics with $m\boldsymbol{k}_0$ and the backscattered Stokes components with $\boldsymbol{k}'_n$ do result in waves whose wave vectors are directed along the propagation axis, their magnitudes making $(m-n)k_0 + k_\mathrm{p}$. Due to this, the scattering pattern looks like there are anti-Stokes components in the slab geometry case as well. For this reason, anti-Stokes scattering components in the three-dimensional case will be observed in narrow solid angles near the pump electromagnetic wave's propagation axis.

The increase in the growth rate at $|\boldsymbol{k}| \to \infty$ corresponds to the generation of a continuum of scattered radiation, i.e., to the emergence of radiation that has a continuous spectrum. It is well known that synchrotron radiation is emitted by an electron traveling along a circular orbit and this radiation comprises an infinite set of harmonics [32]. When circular orbits are distorted, the radiation spectrum changes, and the continuum emerges. There are at least three additional reasons for continuum generation in experiments:

- Bremsstrahlung and, partially, photo-recombination radiation in the plasma.
- The laser radiation is nonmonochromatic, which is of particular importance for ultrashort pulses (the role of this particular circumstance in the propagation of a linearly polarized electromagnetic wave in a plasma will be investigated in the next chapter).
- The nonharmonic character of the electron currents in plasma. This circumstance is related to the onset of plasma turbulence and the anomalous increase in the emission of radiation from it.

Eigenvectors of set (4.11) were calculated, as well as corresponding growth rates. In slab geometry, the eigenvector corresponding to the maximal growth rate has a resonant character, i.e., it is not equal to zero for those values of χ that lie in the vicinity of the harmonics wave vector. In the interval between the nearest two harmonics wave vectors, only the components of the eigenvectors corresponding to these two harmonics are not equal to zero. It is of interest that in slab geometry, the eigenvectors are found as superpositions of left and right circular polarizations.

A large number of eigenvector components are not equal to zero in the continuum area for scattering at large angles to the propagation direction. These numerical data are not informative since the eigenvectors cannot be normalized numerically.

The analysis of slab geometry problem makes it possible to conclude that an initial perturbation which is localized in the $\boldsymbol{k}$-space evolves generating the nearest harmonics consequently.

The present theory leads to the following hypothesis on the polarization of scattered radiation. Since at each moment, perturbations in infinitesimal volumes of plasma are sums of perturbations generated at different times during the pump electromagnetic wave period (the perturbation problem is linearized), and the asymptotic solutions corresponding to different initial moments of time differ from each other by rotation of the $\boldsymbol{k}$-space by an azimuthal phase angle α around the $\hat{\mathbf{e}}_3$ axis, one should expect that the resulting vector potential average values satisfy the following relations: $\langle A_1 \rangle = 0$, $\langle A_2 \rangle = 0$, $\langle A_3 \rangle \neq 0$. Naturally the average values of the squares of all of these quantities are not equal to zero. Thus the radiation should be partially depolarized. A more detailed treatment of these issues must be based on the methods of statistical physics [146] and extends beyond the scope of this study.

The above general theory is applicable in the nonrelativistic limit as well. In this case, the terms proportional to a_0^2 should be dropped, and $\gamma_0 = 1$ should be assumed in (4.1)–(4.4). In contrast to the relativistic case, the corresponding eigenvalue problem is posed for a 15-diagonal matrix. The nonrelativistic simulations for $a_0 = 0.1$ yield results that are identical to those illustrated by Fig. 4.4.

Nonrelativistic problems have been examined for a long time, and typical approaches used were (a) the study of instabilities with the help of resonance approximations based on exact phase matching [144] and (b) treatment of the dispersion relations without phase matching conditions [145] (initial and boundary-value problems). As follows from this study, resonance approximations should be avoided in the relativistic case due to relativistic broadening of resonant structures.

4.4 Conclusions

The results of a linear three-dimensional instability analysis for the propagation of relativistically intense plane monochromatic circularly polarized electromagnetic waves in cold underdense plasmas are presented. The following phenomena are described by the theory:

- excitation of the propagating laser radiation harmonics in the nonlinear medium;
- electromagnetic radiation stimulated scattering by plasmons;
- scattering due to the fluid dynamics analog of Compton effect;
- decay instability at harmonics resulting in scattered electromagnetic waves and plasmons;
- interactions of electromagnetic waves in a plasma;
- generation of a scattered radiation continuum.

Computations show that both forward and backward scattering is possible. Radiation comprises a set of harmonics. The harmonics are scattered into a set of spatial cones embedded in one another. Higher harmonics are scattered into narrower cones. The spectrum of the radiation scattered outside the cones is continuous; the latter phenomenon plays the predominant role.

Harmonics with lower numbers which are scattered into relatively wide cones can propagate outside the spatial area in which the pump laser pulse is localized, whereas the high number harmonics propagate with the pulse. It should be possible to record them experimentally by pulse spectral analysis, using specially arranged geometry. The backscattering efficiency is low due to the short time of interaction between counterpropagating waves [121].

The slab geometry theory of relativistically intense laser radiation scattering in plasmas developed without assuming conservation of the pump laser pulse polarization is a partial case of the general theory discussed in this chapter. In this case, scattering results in the excitation of a sequence of harmonics. Every harmonic is a doublet consisting of Stokes and anti-Stokes components. The relativistic and charge-displacement nonlinearities are the mechanisms of harmonic excitation.

It is established in the framework of the three-dimensional theory that the most significant effects of scattering at large angles to the propagation direction are the generation of a Stokes component at the laser pulse frequency and of the scattered radiation continuum. These phenomena can be observed experimentally.

The harmonics originating from scattering into small solid angles include both Stokes and anti-Stokes components; the latter result from the interactions of higher order harmonics and backscattered Stokes components. The harmonics are doublets when the pump laser pulse intensity is subrelativistic, but components become broader and merge in the case of relativistically intense pump laser radiation. The latter fact can be helpful in experimental studies of laser–plasma interactions since the dependence of the scattering bandwidths on the laser intensity can be used in the propagation process diagnostics.

The predictions of the three-dimensional theory for scattering angles equal to 0 and π are identical to slab geometry results.

The technique of calculating maximal growth rates numerically as the eigenvalues of the matrix of the right-hand sides of an infinite set of coupled ordinary differential equations is a distinctive feature of the approach used for the class of problems treated. It appears that this approach is more efficient than deriving and investigating extremely complicated dispersion relations for growth rates.

The following facts are important for the comparison of the proposed theoretical results with experimental data. First, absorption of laser radiation in a plasma is neglected in the calculations. Including absorption in the model can alter the results slightly. Second, particular experimental geometry and the impact of a finite pulse duration are important. The theory presented above is applicable if (a) the minimal scattering domain (the beam transverse size) is much greater than the radiation wavelength; (b) the development time for the instability is shorter than the pulse duration.

5. Instabilities of Linearly Polarized Plane Electromagnetic Waves in Plasmas

This chapter is dedicated to the instabilities of linearly polarized relativistically intense electromagnetic radiation in cold underdense plasmas. The corresponding linearly polarized electromagnetic waves (or, more precisely, the waveforms comprising an electromagnetic wave and a plasma wave) were described numerically and analytically in Chap. 3. The method of treating the above instability is basically the same as that applied in the previous chapter: once again, below, we will deal with linearized equations with oscillating coefficients which are related to an infinite set of coupled ordinary differential equations in the Fourier space, and the growth rate of instability can be found as the eigenvalue of this set's matrix, which is approximated by a sufficiently large finite-dimensional one. But the physics of the linearly polarized intense electromagnetic radiation instability is different from that of circularly polarized monochromatic electromagnetic waves. As shown in this chapter, the plasma's dielectric response modulation induced by an intense monochromatic linearly polarized electromagnetic wave results in the formation of a plasma analog of a grid, and diffraction on it is described by an analog of Bragg's law.

5.1 3-D Instability Equations

Let us return to linearized equations (4.1)–(4.4). Since our goal in this chapter is to develop a theory of linearly polarized relativistically intense electromagnetic radiation in a cold underdense plasma, it would be appropriate to use the Akhiezer–Polovin problem solutions depicted in Figs. 3.1–3.3 and represented approximately by (3.17) and (3.18) as the ground states. Denote the coefficients of linearized equations (4.1)–(4.4), which depend on the solutions of the Akhiezer–Polovin problem, as follows:

$$f^1(\xi) = \frac{n_0(\xi)}{\gamma_0(\xi)}, \qquad f^2(\xi) = \frac{A_{1,0}(\xi)}{\gamma_0(\xi)}, \qquad f^3(\xi) = \frac{P_{3,0}(\xi)}{\gamma_0(\xi)},$$
$$f^4(\xi) = \frac{n_0(\xi)A_{1,0}(\xi)}{\gamma_0^2(\xi)}, \qquad f^5(\xi) = \frac{n_0(\xi)P_{3,0}(\xi)}{\gamma_0^2(\xi)}.$$

Since the above functions are periodic, they can be expanded in Fourier series

$$f^j(\xi) = \sum_m f^j_m \exp(imk_\mathrm{p}\xi) . \tag{5.1}$$

Obviously, it follows from the above definitions that the Fourier coefficients are related by

$$f^4_m = \sum_j f^1_j f^2_{m-j} , \qquad f^5_m = \sum_j f^1_j f^3_{m-j} .$$

Substituting these series in linearized equations (4.1)–(4.4) and applying the Fourier transform (4.5) $[\boldsymbol{k} = (k_1, k_2, \chi)]$, as we did in the previous chapter for circularly polarized pump electromagnetic waves, we arrive at the following set of equations describing the relativistically intense electromagnetic wave instability in a cold underdense plasma:

$$\begin{aligned} -\left(|\boldsymbol{k}|^2 + \hat{D}_t^2\right) \delta A_{1_0} = \sum_m (&a^1_m \delta A_{1_{\chi - mk_\mathrm{p}}} + a^2_m \delta A_{2_{\chi - mk_\mathrm{p}}} \\ &+ a^3_m \delta\psi_{\chi - mk_\mathrm{p}} + a^4_m \delta n_{\chi - mk_\mathrm{p}}) , \end{aligned} \tag{5.2}$$

$$\begin{aligned} -\left(|\boldsymbol{k}|^2 + \hat{D}_t^2\right) \delta A_{2_0} = \sum_m (&b^1_m \delta A_{1_{\chi - mk_\mathrm{p}}} + b^2_m \delta A_{2_{\chi - mk_\mathrm{p}}} \\ &+ b^3_m \delta\psi_{\chi - mk_\mathrm{p}} + b^4_m \delta n_{\chi - mk_\mathrm{p}}) , \end{aligned} \tag{5.3}$$

$$\begin{aligned} \hat{D}_t \delta n_0 = \sum_m (&c^1_m \delta A_{1_{\chi - mk_\mathrm{p}}} + c^2_m \delta A_{2_{\chi - mk_\mathrm{p}}} \\ &+ c^3_m \delta\psi_{\chi - mk_\mathrm{p}} + c^4_m \delta n_{\chi - mk_\mathrm{p}}) , \end{aligned} \tag{5.4}$$

$$\begin{aligned} \hat{D}_t \delta\psi_0 = \sum_m (&d^1_m \delta A_{1_{\chi - mk_\mathrm{p}}} + d^2_m \delta A_{2_{\chi - mk_\mathrm{p}}} \\ &+ d^3_m \delta\psi_{\chi - mk_\mathrm{p}} + d^4_m \delta n_{\chi - mk_\mathrm{p}}) , \end{aligned} \tag{5.5}$$

where

$$\begin{aligned} a^1_m = &\left(1 + \frac{k_1^2 m k_\mathrm{p}}{|\boldsymbol{k}|^2(\chi - mk_\mathrm{p})}\right) f^1_m - \sum_l \left[\left(1 - \frac{k_1^2}{|\boldsymbol{k}|^2}\right) f^4_l - \frac{k_1\chi}{|\boldsymbol{k}|^2} f^5_l\right] \\ &\times \left(f^2_{m-l} - \frac{k_1}{\chi - mk_\mathrm{p}} f^3_{m-l}\right) , \end{aligned}$$

$$\begin{aligned} a^2_m = &\frac{k_1 k_2 m k_\mathrm{p}}{|\boldsymbol{k}|^2(\chi - mk_\mathrm{p})} f^1_m + \sum_l \left[\left(1 - \frac{k_1^2}{|\boldsymbol{k}|^2}\right) f^4_l - \frac{k_1\chi}{|\boldsymbol{k}|^2} f^5_l\right] \\ &\times \frac{k_2}{\chi - mk_\mathrm{p}} f^3_{m-l} , \end{aligned}$$

$$\begin{aligned} a^3_m = &\frac{\mathrm{i} k_1 \chi m k_\mathrm{p}}{|\boldsymbol{k}|^2} f^1_m - \mathrm{i} \sum_l \left[\left(1 - \frac{k_1^2}{|\boldsymbol{k}|^2}\right) f^4_l - \frac{k_1\chi}{|\boldsymbol{k}|^2} f^5_l\right] \\ &\times \left(k_1 f^2_{m-l} + (\chi - mk_\mathrm{p}) f^3_{m-l}\right) , \end{aligned}$$

$$a_m^4 = f_m^2 - \frac{k_1}{|\boldsymbol{k}|^2}(k_1 f_m^2 + \chi f_m^3),$$

$$b_m^1 = \frac{k_1 k_2 m k_\mathrm{p}}{|\boldsymbol{k}|^2(\chi - m k_\mathrm{p})} f_m^1 + \sum_l \frac{k_2}{|\boldsymbol{k}|^2}\left(k_1 f_l^4 + \chi f_l^5\right) \times \left(f_{m-l}^2 - \frac{k_1}{\chi - m k_\mathrm{p}} f_{m-l}^3\right),$$

$$b_m^2 = \left[1 - \frac{k_2^2 m k_\mathrm{p}}{|\boldsymbol{k}|^2(\chi - m k_\mathrm{p})}\right] f_m^1 - \sum_l \frac{k_2^2}{|\boldsymbol{k}|^2(\chi - m k_\mathrm{p})} \times \left(k_1 f_l^4 + \chi f_l^5\right) f_{m-l}^3,$$

$$b_m^3 = \mathrm{i}\frac{k_2 \chi m k_\mathrm{p}}{|\boldsymbol{k}|^2} f_m^1 - \mathrm{i}\frac{k_2}{|\boldsymbol{k}|^2}\sum_l \left(k_1 f_l^4 + \chi f_l^5\right) \times \left[k_1 f_{m-l}^2 + (\chi - m k_\mathrm{p}) f_{m-l}^3\right],$$

$$b_m^4 = -\frac{k_2}{|\boldsymbol{k}|^2}(k_1 f_m^2 + \chi f_m^3),$$

$$c_m^1 = \mathrm{i}\frac{k_1 m k_\mathrm{p}}{\chi - m k_\mathrm{p}} f_m^1 - \mathrm{i}\sum_l \left(k_1 f_l^4 + \chi f_l^5\right)\left(f_{m-l}^2 - \frac{k_1}{\chi - m k_\mathrm{p}} f_{m-l}^3\right),$$

$$c_m^2 = \mathrm{i}\frac{k_2 m k_\mathrm{p}}{\chi - m k_\mathrm{p}} f_m^1 + \mathrm{i}\sum_l \frac{k_2}{\chi - m k_\mathrm{p}}\left(k_1 f_l^4 + \chi f_l^5\right) f_{m-l}^3,$$

$$c_m^3 = (|\mathbf{k}|^2 - \chi m k_\mathrm{p}) f_m^1 - \sum_l \left(k_1 f_l^4 + \chi f_l^5\right) \times \left[k_1 f_{m-l}^2 + (\chi - m k_\mathrm{p}) f_{m-l}^3\right],$$

$$c_m^4 = -\mathrm{i}(k_1 f_m^2 + \chi f_m^3),$$

$$d_m^1 = -f_m^2 + \frac{k_1}{\chi - m k_\mathrm{p}} f_m^3, \qquad d_m^2 = \frac{k_2}{\chi - m k_\mathrm{p}} f_m^3,$$

$$d_m^3 = -\mathrm{i}\left[k_1 f_m^2 + (\chi - m k_\mathrm{p}) f_m^3\right], \qquad d_m^4 = -|\mathbf{k}|^{-2}\delta_{m,0}.$$

The above set of ordinary differential equations can be put in the standard form given by (4.11) [67]. Once again, the corresponding matrix is denoted by B. The equation for this matrix eigenvalues is $det|B - G_m I| = 0$. Obviously, it is invariant with the respect to the shift of the argument χ by k_p. For this reason, the matrix B eigenvalues $G_m = \mathrm{Re}G_m + i\mathrm{Im}G_m$ are periodic in χ; the period is k_p. Similarly to the previous chapter, the growth rate of instability is defined as the maximal positive real part of the matrix B eigenvalue, $G = \max \mathrm{Re}G_m$.

5.2 Scattering of Linearly Polarized Electromagnetic Waves in 1-D Geometry

The 1-D case of relativistically intense electromagnetic wave instability in a cold plasma was investigated in [67]. The equations to describe it are obtained from those of the previous section by setting $k_1 = k_2 = 0$. As shown in this work in the 1-D case, linearly polarized electromagnetic wave instability includes both forward and backward scattering. The perturbation spectrum comprises a periodic set of bands corresponding to stimulated Raman scattering harmonics; the spacing between the bands is k_p. The specific shapes of bands are also determined by the fluid dynamics analog of Compton scattering of photons by electrons which are driven by the pump wave at velocities comparable to the speed of light. Langmuir noise generation by the propagating pump wave plays a major role in forming the edges of the bands.

A comparison of the instabilities of plane linearly polarized (nonmonochromatic) and circularly polarized (monochromatic) waves [44] demonstrates that the qualitative character of the growth rates in these two cases is substantially different. The harmonics spacing is k_0 for circular polarization and k_p for linear polarization. In the 1-D case, the impact of Compton scattering is insignificant for circular polarization of the pump electromagnetic wave, but it is this effect that determines the band center shape for linear polarization. In the subrelativistic intensity range, the growth rate of circular polarization instability clearly includes Stokes and anti-Stokes components, which is another distinction from linear polarization instability.

5.2.1 One-Dimensional Scattering Equations

Assuming $k_1 = k_2 = 0$, in slab geometry, from (5.2)–(5.5),

$$-\left(\chi^2 + \hat{D}_t^2\right) \delta A_{1_0} = \sum_m (a_m^1 \delta A_{1_{\chi - m k_\mathrm{p}}} + a_m^2 \delta A_{2_{\chi - m k_\mathrm{p}}} + a_m^3 \delta\psi_{\chi - m k_\mathrm{p}} + a_m^4 \delta n_{\chi - m k_\mathrm{p}}) , \tag{5.6}$$

$$-\left(\chi^2 + \hat{D}_t^2\right) \delta A_{2_0} = \sum_m (b_m^1 \delta A_{1_{\chi - m k_\mathrm{p}}} + b_m^2 \delta A_{2_{\chi - m k_\mathrm{p}}} + b_m^3 \delta\psi_{\chi - m k_\mathrm{p}} + b_m^4 \delta n_{\chi - m k_\mathrm{p}}) , \tag{5.7}$$

$$\hat{D}_t \delta n_0 = \sum_m (c_m^1 \delta A_{1_{\chi - m k_\mathrm{p}}} + c_m^2 \delta A_{2_{\chi - m k_\mathrm{p}}} + c_m^3 \delta\psi_{\chi - m k_\mathrm{p}} + c_m^4 \delta n_{\chi - m k_\mathrm{p}}) , \tag{5.8}$$

$$\hat{D}_t \delta\psi_0 = \sum_m (d_m^1 \delta A_{1_{\chi - m k_\mathrm{p}}} + d_m^2 \delta A_{2_{\chi - m k_\mathrm{p}}} + d_m^3 \delta\psi_{\chi - m k_\mathrm{p}} + d_m^4 \delta n_{\chi - m k_\mathrm{p}}) . \tag{5.9}$$

The coefficients of the set of equations (5.6)–(5.9) are given by the following relations:

$$a_m^1 = f_m^1 - \sum_l f_{m-l}^2 f_l^4\,, \qquad a_m^2 = 0\,,$$

$$a_m^3 = -\mathrm{i}\sum_l (\chi - mk_\mathrm{p}) f_{m-l}^3 f_l^4\,, \qquad a_m^4 = f_m^2\,,$$

$$b_m^1 = 0\,, \qquad b_m^2 = f_m^1\,, \qquad b_m^3 = 0\,, \qquad b_m^4 = 0\,,$$

$$c_m^1 = \mathrm{i}\chi \sum_l f_l^5 f_{m-l}^2\,, \qquad c_m^2 = 0\,,$$

$$c_m^3 = \chi(\chi - mk_\mathrm{p})(f_m^1 - \sum_l f_l^5 f_{m-l}^3)\,, \qquad c_m^4 = -\mathrm{i}(k_1 f_m^2 + \chi f_m^3)\,,$$

$$d_m^1 = -f_m^2\,, \qquad d_m^2 = 0\,,$$

$$d_m^3 = -\mathrm{i}(\chi - mk_\mathrm{p}) f_m^3\,, \qquad d_m^4 = -\chi^{-2}\delta_{m,0}\,.$$

5.2.2 Propagation of Perturbations Parallel to the Pump Wave

The set of equations (5.6)–(5.9), together with the above coefficient definitions, can be divided into two blocks for the functions δA_1, $\delta\psi$, δn and δA_2 respectively. The main interest lies in the propagation of a perturbation δA_1 polarized parallel to the reference wave.

In the framework of the approach adopted, the dependence of the growth rate $G(\chi)$ on the wave vector longitudinal component χ (which is the argument of the Fourier transform) must be found from a set of differential equations with periodic coefficients. This technique makes it possible to consider an ensemble of instabilities, including superintense laser radiation stimulated Raman scattering by plasmons, its self-modulation, and interactions between electromagnetic waves.

We analyzed the problem by selecting the perturbative analysis ground state solutions with average normalized intensities in the $\overline{I} = 0.1$–400 range. These solutions were obtained by varying the "initial conditions" for the Akhiezer–Polovin waves, one of which is shown in Fig. 3.1.

The initial linearized instability model (4.1)–(4.4) represents a set of linear partial differential equations with periodic coefficients. Transition to the spatial Fourier transforms leads to an infinite linear set of ordinary differential equations in time. The growth rate is defined as the maximum eigenvalue of an infinite-dimensional matrix of this system. Similarly to the previous chapter, to perform calculations, the infinite-dimensional matrix is approximated by a square matrix of finite rank. Estimates indicate that this matrix rank should be at least 300. Again, the growth rate is found with the help of the QR algorithm described in [143].

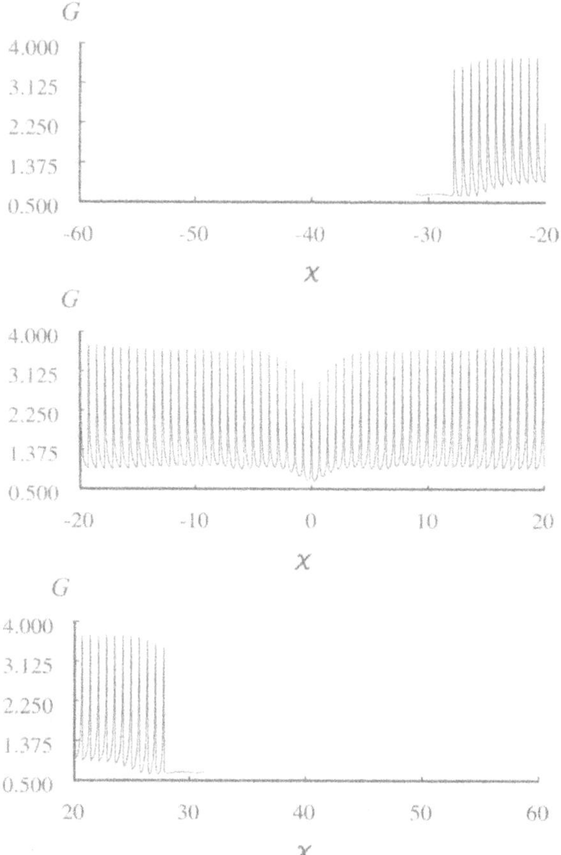

Fig. 5.1. Slab geometry growth rate G for the ground state electromagnetic wave depicted in Fig. 3.1. The dimension of matrix B used is 316 [121]

Figures 5.1 and 5.2 show the results of growth rate calculations found by using square matrices having a gradually increasing rank $R = 4(2j + 1) = 316, 604$. The corresponding number of harmonics taken into account in the Fourier transform space is $2j + 1 = 79, 151$.

The growth rate is symmetrical with respect to χ. Calculations demonstrate the existence of an extensive boundary effect, which affects over 40 harmonics for positive and negative values of χ. The correct result is reflected by the central part of the growth rate distribution for the matrix of rank 604, which represents a periodic structure of lines shifted relative to one another by k_{p}. An increase in the rank of the matrix used in the calculations yields a larger correct central (periodic) part of the growth rate distribution and affects an ever increasing number of harmonics.

The problem is thus associated with finding the solution for a single period k_{p}, or describing one line. The difficulty lies in the fact that a large number

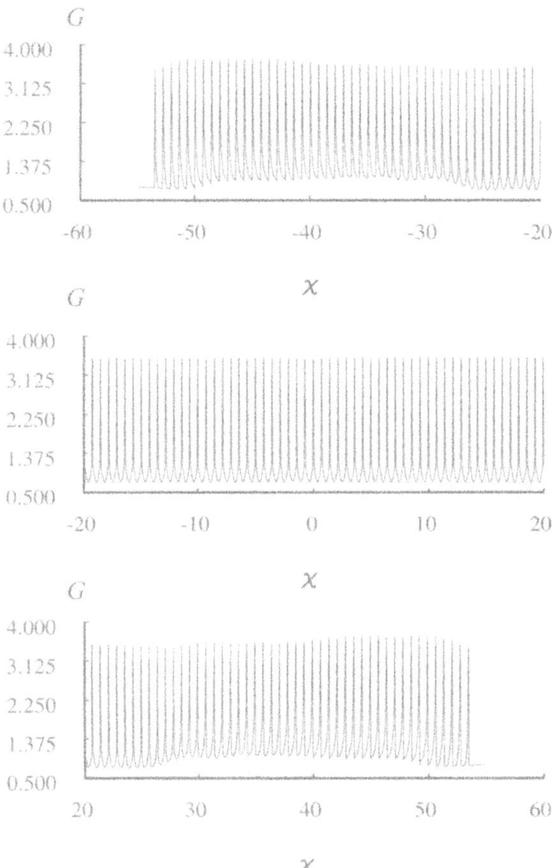

Fig. 5.2. Slab geometry growth rate G for the ground state electromagnetic wave depicted in Fig. 3.1. The dimension of matrix B used is 604 [121]

of harmonics contributes to the solution for any one line, making it necessary to use high-rank, finite-dimensional matrices in computations.

The first three eigenvalues for one line are depicted in Fig. 5.3. Since, in accord with the adopted definition, the growth rate is the maximal real part of all of the problem eigenvalues, one can see in this figure that, at different values of χ, different eigenvalues are "responsible" for the growth rate. There is a singularity at the line center originating from the singularities in the coefficients of (5.6)–(5.9), whose appearance is attributed to the presence of long-wavelength perturbations characterized by $\chi \to 0$ in the adopted model. Under experimental conditions the perturbation wavelength is limited, for example, by plasma dimensions, the laser pulse length, and the presence of absorption in the propagation medium. Therefore, long-wavelength perturbations should be excluded from consideration. A potential technique for this

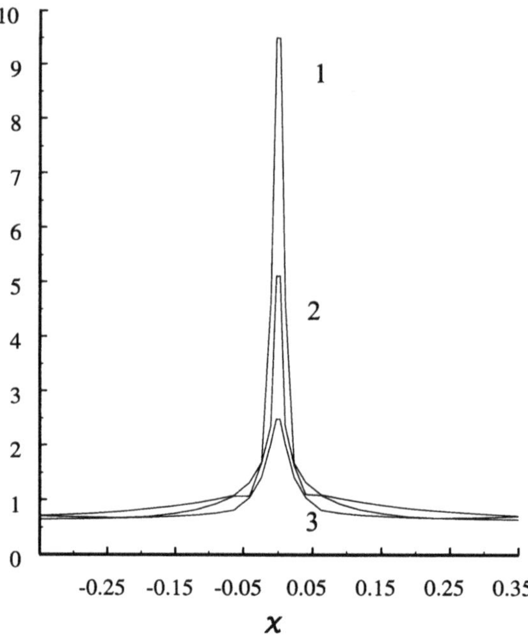

Fig. 5.3. The first three eigenvalues for the instability of the ground state electromagnetic wave depicted in Fig. 3.1 [121]

purpose is to impose the condition that $\delta n_0 \to 0$ in the $\chi \to 0$ limit. The integral of the growth-rate line profile converges and this makes it possible to introduce correctly the concept of an integral bandwidth. All of the growth rate calculations described here were performed excluding regions in the vicinity of the singularities. A change of the eigenvalue solution branches occurs within the wings of the line.

Figure 5.4 illustrates the instability growth rate lines for various intensities. The separation between adjacent lines depends on the pump wave intensity.

We calculated the eigenvectors of the problem considered as well, which made it possible to identify the nature of perturbations in various spectral parts. The δA_1, $\delta\psi$, and δn perturbations are generated near the line maximum, and they give rise to a scattered wave and to Langmuir noise. Generation of δA_1 and $\delta\psi$ occurs at some distance from the line center. This effect is due to the fluid dynamics analog of relativistic Compton scattering of pump electromagnetic wave photons by moving electrons. The δA_1, $\delta\psi$, and δn perturbations, which correspond to the interaction with plasmons, emerge again in the line edges. The value of χ is arbitrary for plasmons in the cold plasma model and, therefore, far away at the line edges, the growth rate depends on it weakly (white noise).

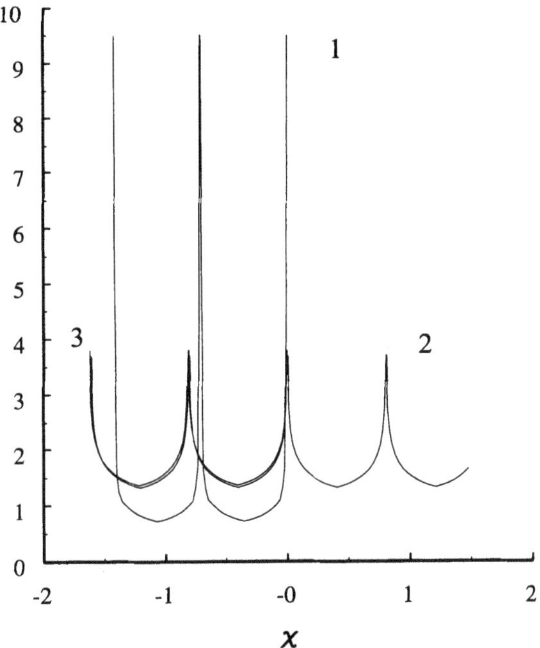

Fig. 5.4. Instability growth rates for the Akhiezer–Polovin waves with the normalized intensities making (1) 51.5, (2) 3.82, and (3) 0.11 [121]

5.3 Scattering Diagrams for 3-D Instability

Consider the growth rate for problem (5.2)–(5.5) as a function of the three components of the perturbation wave vector **k**. The growth rate distribution can be presented graphically as a function of two variables in Cartesian or spherical coordinates. In Cartesian coordinates, the superintense electromagnetic pump wave propagates in the direction of the $\boldsymbol{e}_3$ axis and is polarized in the $\boldsymbol{e}_1$ direction. Let θ and α be the polar and azimuthal angles in $\boldsymbol{k}$ space. The growth rate of relativistically intense linearly polarized electromagnetic wave instability is shown in Figs. 5.5 and 5.6 for $k_2 = 0$.

Now let us consider the growth rate distribution as a function of θ and α for fixed values of $k = |\boldsymbol{k}|$. The latter correspond to specific values of the scattered radiation frequency roughly given by $\omega = k/c$, whereas the θ and α angles define the scattering direction. Since measurements in experiments are performed at various angles, the analysis of spectral data at specific frequencies should make it possible to use the dependencies presented below for interpreting experimental data. The growth rate distribution G is depicted in Figs. 5.7–5.10 as a function of $(\cos\theta, \alpha)$ (the corresponding pump waves are shown in Figs. 3.1–3.3). The propagating radiation is scattered into a countable set of embedded cones. The scattering pattern is independent of the azimuthal angle within each cone. Scattering is symmetrical with respect to

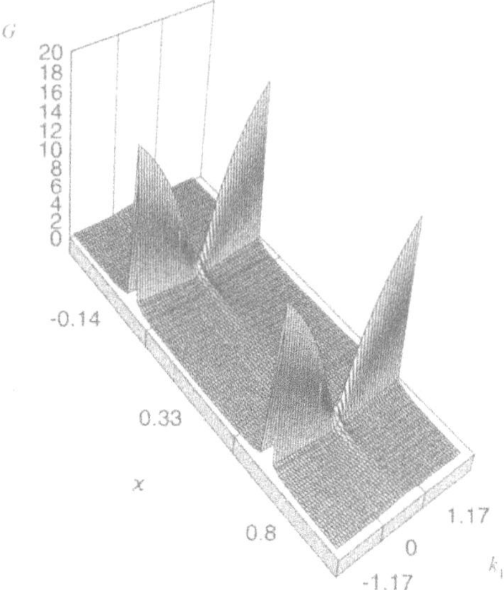

Fig. 5.5. The dependence of growth rate G on (k_1, χ) for $k_2 = 0$. The ground state electromagnetic wave is depicted in Fig. 3.1. Singularities at the symmetry axis resulting from the fluid dynamics analog of the Compton effect are excluded

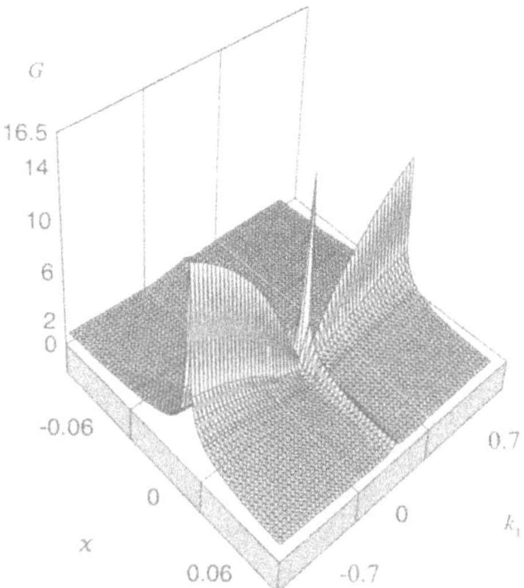

Fig. 5.6. The dependence of growth rate G on (k_1, χ) for $k_2 = 0$. The singularity at the center is due to the fluid dynamics analog of the Compton effect. The ground state electromagnetic wave is depicted in Fig. 3.1

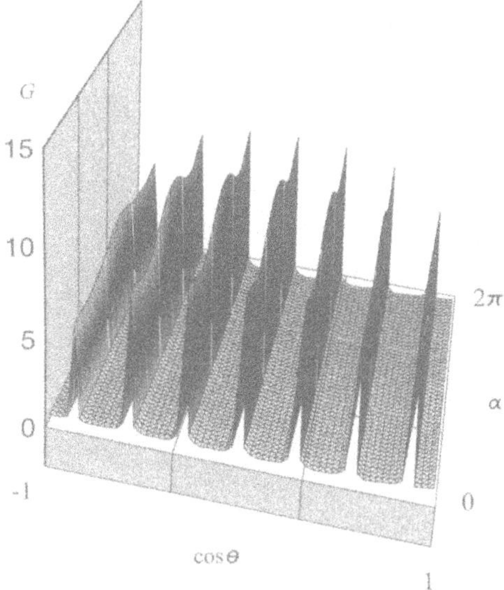

Fig. 5.7. The dependence of growth rate G on $(\cos\theta, \alpha)$ for $\chi = 3.5\, k_{\mathrm{p}}$. The ground state electromagnetic wave is depicted in Fig. 3.1

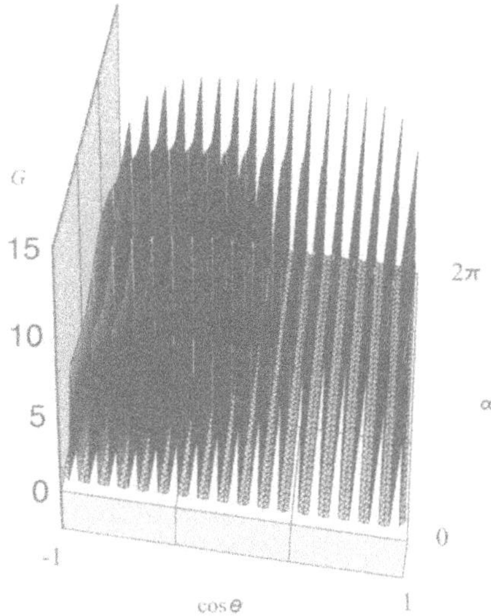

Fig. 5.8. The dependence of growth rate G on $(\cos\theta, \alpha)$ for $\chi = 8.5\, k_{\mathrm{p}}$. The ground state electromagnetic wave is depicted in Fig. 3.1

$\theta = \pi/2$ (backward and forward scattering) and with respect to the polarization plane $\alpha = 0$. Obviously, due to the relativistic nonlinearity saturation, the growth rates of more intense pump electromagnetic waves tend to be lower.

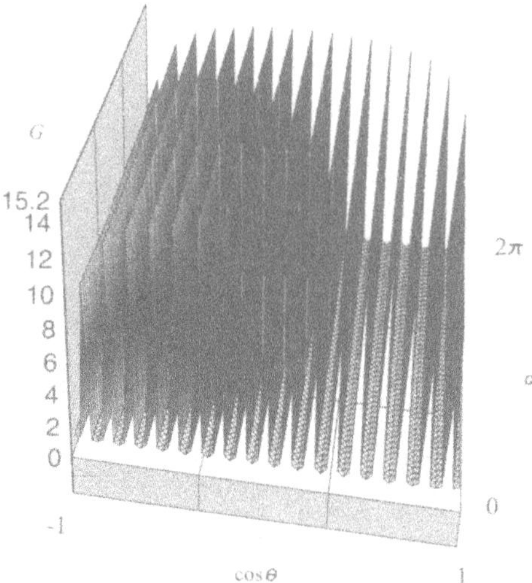

Fig. 5.9. The dependence of growth rate G on $(\cos\theta, \alpha)$ for $\chi = 8.5\, k_p$. The ground state electromagnetic wave is depicted in Fig. 3.2

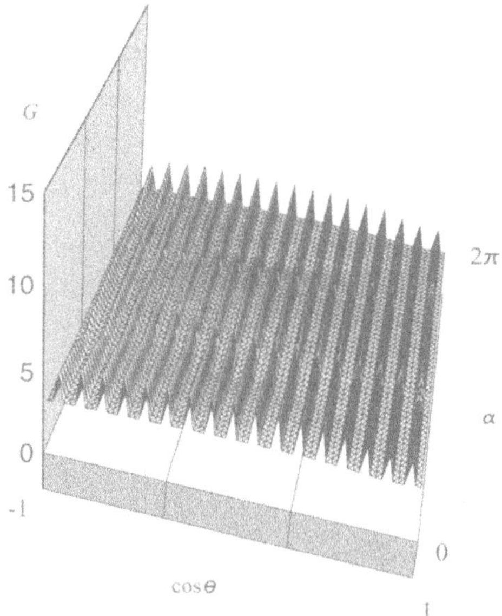

Fig. 5.10. The dependence of growth rate G on $(\cos\theta, \alpha)$ for $\chi = 8.5\, k_{\mathrm{p}}$. The ground state electromagnetic wave is depicted in Fig. 3.3

The dependence of G on α (in polar coordinates) for the two scattering cones $\cos\theta = mk_{\mathrm{p}}/k$, where $m = 1, 8$ and $k = 8.5\, k_{\mathrm{p}}$, is shown in Fig. 5.11. An important feature illustrated by this diagram is the scatter-

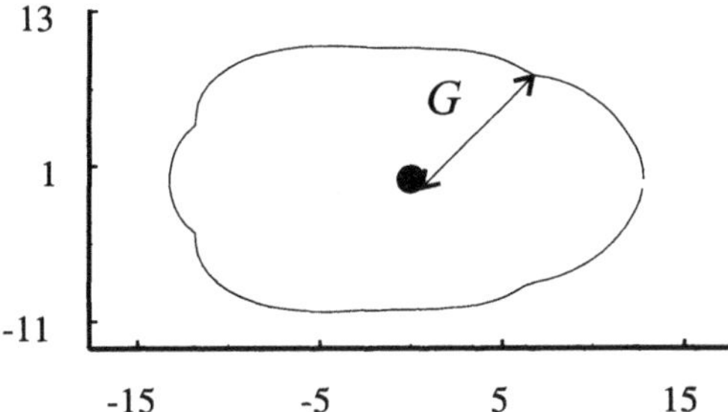

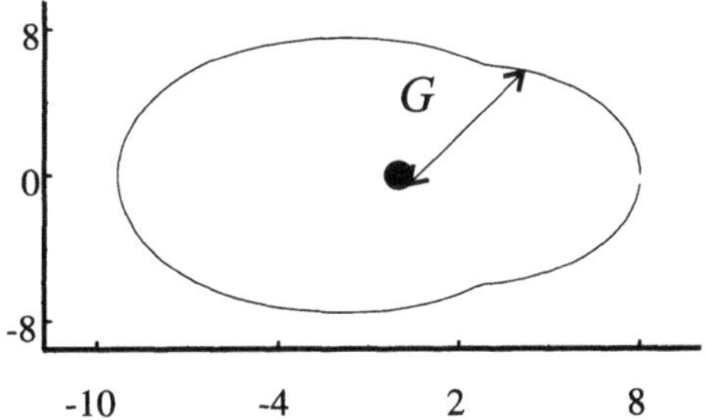

Fig. 5.11. The dependence of G on α (in polar coordinates) for the two scattering cones given by $\cos\theta = mk_{\mathrm{p}}/k$, where $m = 1, 8$ and $k = 8.5\,k_{\mathrm{p}}$. The ground state electromagnetic wave is depicted in Fig. 3.1

ing polar anisotropy. As one can see in this figure, scattering at $\alpha = 0$ and π (in the pump wave polarization direction) prevails.

The growth rate distributions G as functions of $\cos\theta$ for $\alpha = 0$ and for a range of values of k/k_{p} are given in Fig. 5.12. The transformation of the scattering angular diagram structure for a sequence of increasing values of k (for increasing scattered radiation frequencies) is shown in this figure. Narrow scattered radiation cones directed at $\theta = 0$ and π (forward and backward) emerge for integer values of k/k_{p} (see Fig. 5.12c). This fact is explained by the fluid dynamics analog of Compton scattering. Further increase in the value of k results in moving the emerging maxima closer to the center and in the emergence of a pair of new local maxima each time k/k_{p} is an integer number. Thus, we see eight pairs of local maxima and a central one in Fig. 5.12c.

The coordinates of the maxima are given by

$$\cos\theta = mk_{\mathrm{p}}/k\,, \qquad m = 0, \pm 1, \pm 2, \dots .$$

This relation allows a simple interpretation. As one can see in Fig. 3.1, an intense electromagnetic wave propagating into a plasma induces a periodic modulation of the medium's dielectric response. This modulated medium can be roughly approximated by a periodic grid. In this framework, the above

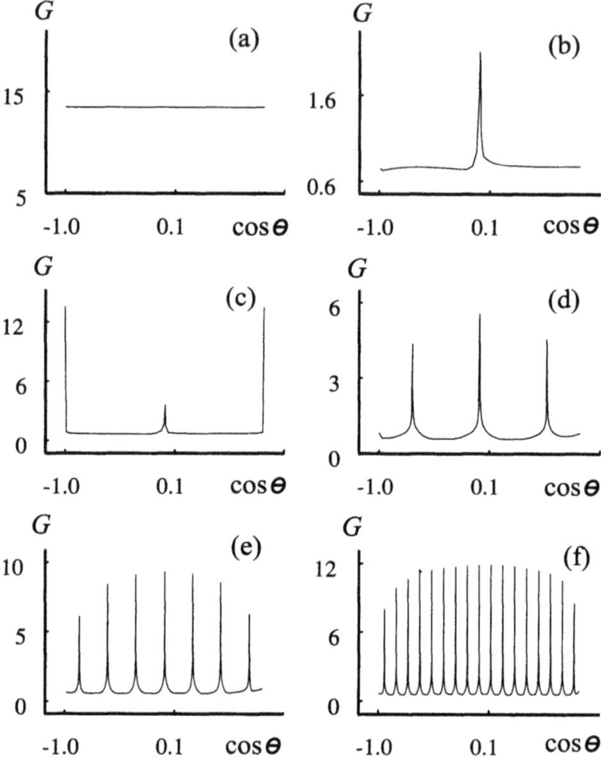

Fig. 5.12. The dependence of growth rate G on $\cos\theta$ for the azimuthal angle $\alpha = 0$, which corresponds to the maximal scattering efficiency. The values of the parameter k/k_{p} are (**a**) 0; (**b**) 0.5; (**c**) 1; (**d**) 1.5; (**e**) 3.5; (**f**) 8.5

equation is an analog of the Bragg diffraction law with a factor of 2 correction resulting from the Doppler shift associated with the use of a co-moving variable.

Note that the growth rate structure includes a continuous background corresponding to the emergence of a continuum of scattered radiation. Separate local maxima resulting from Compton scattering by plasmons and the fluid dynamics analog of Compton scattering are seen against this background.

5.4 Conclusions

The results of a formal linear analysis of the relativistically intense linearly polarized electromagnetic wave propagation instability in cold underdense plasmas are presented in this chapter in the framework of 1-D and 3-D geometries. The nature of the instability is described. As we have seen, the fact that the superintense electromagnetic pump wave is not monochromatic leads to substantial complication of the theory. Physically, the nonmonochromatic electromagnetic wave given by a solution of the Akhiezer–Polovin problem can be interpreted as a flux of photons with frequencies shifted by an integer multiple of the plasma frequency. Such waves are used as ground states for the analysis of propagation instability. The relativistic theory presented in this chapter includes

- excitation of Compton scattering harmonics of relativistically intense electromagnetic pump wave propagating in nonlinear medium;
- the fluid dynamics analog of the Compton effect resulting from momentum exchange with cold plasma electrons;
- interactions of electromagnetic waves in plasmas;
- generation of a continuum of scattered radiation.

The model character of the above theory is due to the "excessive" allowance for arbitrarily long wavelength perturbations, which results in a singularity in the spectral bands of Compton scattering harmonics. This circumstance should not influence the overall scattering picture (the frequencies of scattered waves and their propagation directions, the scattering cones geometry, the scattering anisotropy) substantially. The model used is inapplicable only in small vicinities of the growth rate maxima. Potentially, this problem could be eliminated by developing a theory for finite spatial size pulses and plasmas. In our numerical simulations, adaptive grids were used to treat growth rate singularities.

Initially, the problem of linearly polarized relativistically intense plane wave scattering in a cold underdense plasma was considered on the basis of a spatially one-dimensional model. An important factor in this investigation was selection of a reference wave which is the exact solution of the basic Maxwell equations and the equations describing the relativistic fluid dynamics of electrons. The waves used as the ground states are not monochromatic but represent flows of photons with wave vectors $k_0 \pm lk_\mathrm{p}$ and with frequencies $\omega_0 \pm l\omega_\mathrm{p}$, shifted by $\pm l\omega_\mathrm{p}$, where l is an integer. The frequency ω_p and the wave number $k_\mathrm{p} = \omega_\mathrm{p}/c$ differ from the plasma frequency $\omega_{\mathrm{p}0}$ and from the wave vector $k_{\mathrm{p}0} = \omega_{\mathrm{p}0}/c$ of an unperturbed plasma. There can be various distributions of photon fluxes over the integral numbers m in a plane nonlinear wave [43].

In the relativistic range of intensities, a linearly polarized nonmonochromatic wave is unstable, and this results in forward and backward Raman scattering. Computations show that in the Fourier transform space, the scattering spectrum consists of a periodic set of lines representing harmonics of stimulated Raman scattering, shifted relative to one another by a quantity that is a multiple of k_p. The profile of each specific growth rate line is also influenced by a relativistic fluid dynamics analog of Compton scattering of photons by electrons that are driven by a superintense electromagnetic pump wave at velocities close to the speed of light. The line edges are formed under the influence of Langmuir noise generation by the propagating pump wave.

In principle, an experimental determination of the spacing of the growth rate lines and their widths, for example, for backscattered radiation, should make it possible to estimate the pump wave intensity and the plasma electron density.

It is demonstrated that a plane wave in a plasma is unstable, even in one-dimensional geometry, which can be interpreted as the initial stage of the appearance of electromagnetic turbulence [144].

A comparison of the results of a linear analysis of the propagation instability of strong plane (nonmonochromatic) linearly polarized waves and of circularly polarized (monochromatic) waves (see previous chapter and [44]) in a cold plasma reveals important differences among the overall character of the growth rate lines. There are differences among the growth rate periods separating the adjacent harmonics in the wave vector space (k_0 for circular polarization and k_{p} for linear polarization), as well as in the general line structure. For circular polarization, the influence of the fluid dynamics analog of the Compton effect is unimportant in one-dimensional geometry, whereas in linear polarization this effect is primarily responsible for line center formation. In the nonrelativistic range for circular polarization, the gain line is split into Stokes and anti-Stokes components, whereas for linear polarization, all of the harmonics of stimulated Raman scattering are present.

It was demonstrated by computations that both forward and backward scattering are possible. Radiation scattered in various directions comprises a set of harmonics whose frequencies depend on the scattering angle and propagating against a continuous spectral background. Overall, the scattered radiation spectrum is continuous. At each specific frequency, the electromagnetic radiation is scattered into a countable set of embedded spatial cones with frequency-dependent azimuthal angles.

The scattering growth rate is symmetrical with respect to the pump wave polarization plane and the plane perpendicular to the direction of this wave's propagation. Despite this symmetry, under realistic conditions, the backward scattering efficiency should be limited due to the short interaction time of counterpropagating electromagnetic waves [121]. At the same time, there is an azimuthal anisotropy of the growth rate structure. The prevalent scattering directions lie in the pump wave polarization plane.

As mentioned above, the fluid dynamics analog of Compton scattering is described by the theory presented. This effect plays a major role in forward and backward scattering and affects plasma frequency harmonics. The momentum of an electron moving in a "figure-eight" trajectory in the pump polarization plane is directed parallel to the field momentum at the upper and lower points of its path. Interactions which occur at corresponding times result in the most efficient exchange of momentum between electrons and the electromagnetic field.

Since linearly polarized radiation is used in experiments with relativistic intensities the scattering diagrams presented above can be particularly useful for interpreting experimental data.

6. Models of Nonlinear Propagation of Relativistically Intense Ultrashort Laser Pulses in Plasmas

In this chapter, we examine a number of basic mathematical models of interactions of relativistically intense, ultrashort laser pulses with cold underdense plasmas. The general equations discussed in Chap. 2 are too complicated to treat directly to describe the nonlinear propagation of powerful laser radiation in plasmas (this would entail extremely massive computations and make it hardly possible to derive any analytical results); substantial model simplifications can stem from the fact that in plasmas obtained by ionization of gaseous targets, the laser pulse frequency is much greater than the plasma electron frequency. First, this circumstance makes it possible to average the equations describing the medium's response to the laser pulse electromagnetic field, after which the fluid dynamics equations become much simpler. Second, the traditional nonlinear optics approach of introducing an envelope approximation appears productive in the powerful laser–plasma interactions theory as well. The electromagnetic field's slow amplitude and fast phase can be introduced, assuming that this amplitude varies slowly on the laser radiation wavelength scale. As a result, the evolution of this amplitude is described by a nonlinear Schroedinger equation instead of the wave equation.

However, note that under certain conditions associated with extremely sharp nonlinear self-focusing, the Schroedinger equation model may become inadequate and the terms responsible for high order dispersion must be retrieved. The corresponding "full" models of relativistically intense laser pulse propagation are presented below as well.

The objective of this chapter is to present a hierarchy of models of the interactions between relativistically intense, ultrashort laser pulses and cold underdense plasmas. A general three-dimensional model is considered along with its limits corresponding to large aperture laser pulses and long laser beams. As an illustration of the applications of the models developed, we present results of studies of electromagnetic field instabilities in plasmas corresponding to laser radiation filamentation, self-modulation, stimulated scattering by plasmons, and the generation of third harmonics.

6.1 The Physical Model

In this chapter, nonstationary three-dimensional models describing the nonlinear propagation of circularly polarized, relativistically intense, ultrashort laser pulses in cold underdense plasmas are derived from the Maxwell equations (Coulomb gauge) and the equations of electron fluid dynamics. The most general model incorporates the following phenomena, which were outlined in the Introduction and determine the character of laser–plasma interactions at extremely high intensities:

- laser radiation diffraction and refraction;
- the relativistic nonlinearity which results from the modification of the plasma refractive index due to the increase in the masses of free electrons driven by the laser pulse electromagnetic field at velocities comparable to the speed of light;
- the charge-displacement nonlinearity which stems from the modification of the plasma's refractive index due to the plasma electron component concentration's variation caused by the ponderomotive force;
- excitation of Langmuir waves in plasma by propagating laser radiation;
- electromagnetic wave temporal dispersion in plasmas;
- plasma electron component inertia.

Deriving the model equations, we assume that laser pulse duration is much shorter than the plasma ion component response time and consequently the laser-irradiated plasma ions remain immobile.

Lagrangian formulations and conservation laws for the proposed models are also presented in this chapter. Large pulse aperture and long beam limits are considered. Neglecting high-order laser radiation dispersion, the propagation of long laser beams in cold underdense plasmas is described by the model embodied in the Schroedinger equation with relativistic and charge-displacement nonlinearity, originally proposed in [81]. Detailed studies on the basis of this model are performed in [7,8,95], where a countable set of eigenmodes of this nonlinear Schroedinger equation is presented and their instabilities are studied. It was demonstrated by numerical simulations using the Schroedinger equation with relativistic and charge-displacement nonlinearity that an important regime of confined propagation of ultrashort superintense laser beams exists in matter, called relativistic and charge-displacement self-channeling. The results of these studies are summarized in Chap. 8. It is also shown that in the large aperture limit, the general propagation model developed in this chapter becomes identical to the set of nonlinear equations given in [106].

Note that one of the models considered below is closely related to that proposed in [119]. However, in this paper, a number of significant terms relevant to the model's conservative properties were neglected.

The slow field amplitude approximation was subjected to a critical analysis in a number of works [12, 103, 115–117]. One of the basic conclusions of these papers was that an adequate description of short pulse propagation at large plasma penetration distances in the case of extremely sharp nonlinear self-focusing requires a wave equation whose terms are responsible for high-order dispersion instead of the nonlinear Schroedinger equation. For this reason, contrary to the traditional approach, we do not neglect these terms in some of the models established below.

6.2 Derivation of the Basic Model Equations

Let us adopt the following assumptions:

- The laser radiation is circularly polarized and the vector potential has the form

$$\boldsymbol{A} = \frac{1}{2}(\boldsymbol{e}_1 + \mathrm{i}\boldsymbol{e}_2)a(\boldsymbol{x}_\perp, \eta, \tau)\exp[\mathrm{i}(k_0 x_3 - \omega_0 t)] + c.c.\,, \tag{6.1}$$

  where

$$\tau = \frac{\varepsilon t}{2}\,, \qquad \varepsilon = \omega_0^{-1}\,, \qquad \eta = x_3 - \theta t\,, \tag{6.2}$$

$$\theta = \frac{k_0}{\omega_0}\,, \tag{6.3}$$

$$k_0 = \sqrt{\omega_0^2 - 1}\,. \tag{6.4}$$

  The laser pulse central frequency ω_0 in (6.1)–(6.4) is normalized by ω_p, and ε is considered a small parameter (which means that the plasma is strongly underdense). Equation (6.3) is the definition of the laser pulse group velocity, and dispersion relation (6.4) corresponds to the propagation in an unperturbed plasma. As follows from the Coulomb gauge (2.58) and the condition $\varepsilon \ll 1$, to the lowest order, the vector potential longitudinal component is negligible, and it can be assumed that the propagating electromagnetic radiation remains transversely polarized.
- We assume that the plasma interaction with a powerful ultrashort laser pulse is vortex-free, so that $\alpha \equiv 0$ and $\beta \equiv 0$ in (2.62). Thus, the plasma electron component momentum is

$$\boldsymbol{p} = \boldsymbol{A} + \nabla\psi\,. \tag{6.5}$$

  As mentioned earlier, ψ is the potential of the field–plasma system's generalized momentum. Under this condition, we have one scalar equation,

$$\psi_t = \varphi - \gamma$$

  instead of the Euler equation (2.59).

Introducing the co-moving variable defined by (6.3) and averaging the resulting equations, which is physically equivalent to neglecting the excitation of the propagating laser pulse harmonics, we obtain the following self-consistent set of equations [121]:

$$\mathrm{i}a_\tau + \triangle_\perp a + \varepsilon^2 a_{\eta\eta} + \varepsilon\theta a_{\eta\tau} + \left(1 - \frac{n}{\gamma}\right) a = \frac{\varepsilon^2}{4} a_{\tau\tau}\,, \tag{6.6}$$

$$(\triangle_\perp + \partial_\eta^2)\varphi = n - 1\,, \tag{6.7}$$

$$\gamma - \varphi = \theta\psi_\eta - \frac{\varepsilon}{2}\psi_\tau\,, \tag{6.8}$$

$$\left(\frac{n}{\gamma}\psi_\eta\right)_\eta + (\nabla_\perp, \frac{n}{\gamma}\nabla_\perp\psi) = \theta n_\eta - \frac{\varepsilon}{2} n_\tau\,, \tag{6.9}$$

$$\gamma = \sqrt{1 + |a|^2 + |\nabla_\perp\psi|^2 + \psi_\eta^2}\,. \tag{6.10}$$

Equation (6.6) describes the propagation of ultrashort laser pulses in cold underdense plasmas. Its right-hand term corresponds to high-order dispersion. Equation (6.7) is the Laplace equation in co-moving variables (from the beginning we used the Coulomb gauge). Expressions (6.8) and (6.9) are the averaged generalized momentum and continuity equations. Again, their right-hand sides correspond to the effect of high-order dispersion. The averaged relativistic mass factor is given by (6.10).

One can verify directly that (6.6)–(6.10) can be derived from the following Lagrangian density [121]:

$$\begin{aligned} L = n\left(\frac{\epsilon}{2}\psi_\tau - \theta\psi_\eta + \gamma - \varphi\right) + \varphi - \frac{1}{2}\left(\varphi_\eta^2 + |\nabla_\perp\varphi|^2\right) \\ -\frac{1}{2}\Big(\mathrm{i}a^* a_\tau - (\nabla_\perp a, \nabla_\perp a^*) - \epsilon^2|a_\eta|^2 \\ - \frac{\epsilon\theta}{2}(a_\eta a_\tau^* + c.c.) + \frac{\epsilon^2}{4}|a_\tau|^2 + |a|^2\Big)\,. \end{aligned} \tag{6.11}$$

The invariants of the problem considered are its Hamiltonian,

$$\begin{aligned} H = \int \Big[\epsilon^2|a_\eta|^2 + (\nabla_\perp a, \nabla_\perp a^*) - \varphi_\eta^2 \\ - |\nabla_\perp\varphi|^2 - |a|^2 + 2(\varphi - 1) + \frac{\epsilon^2}{4}|a_\tau|^2 - \epsilon n\psi_\tau\Big]\mathrm{d}^2x_\perp \mathrm{d}\eta\,, \end{aligned} \tag{6.12}$$

field mass

$$N = \int \left[|a|^2 + \frac{\mathrm{i}\varepsilon\theta}{2}(a_\eta^* a - c.c.) + \frac{\mathrm{i}\varepsilon^2}{4}(a^* a_\tau - c.c.)\right] \mathrm{d}^2x_\perp \mathrm{d}\eta\,, \tag{6.13}$$

and momentum

$$\boldsymbol{P}_\perp = \int \Big[\mathrm{i}a^* \nabla_\perp a - \frac{\varepsilon\theta}{2}(a_\eta^* \nabla_\perp a + c.c.) + \frac{\varepsilon^2}{4}(a_\tau^* \nabla_\perp a + c.c.) - \varepsilon n \nabla_\perp \psi \Big] \mathrm{d}^2 x_\perp \mathrm{d}\eta \,, \tag{6.14}$$

$$P_\parallel = \int \Big[\mathrm{i}a^* a_\eta - \varepsilon\theta |a_\eta|^2 (a_\eta^* \nabla_\perp a + c.c.) + \frac{\varepsilon^2}{4}(a_\tau^* a_\eta + c.c.) - \varepsilon n \psi_\eta \Big] \mathrm{d}^2 x_\perp \mathrm{d}\eta \,. \tag{6.15}$$

An attempt to develop a model of laser beam propagation in cold underdense plasmas incorporating the effect of high order dispersion was made in [12, 103, 115–117]. Here, we demonstrate that the model presented and treated in these papers follows from that established above. For laser beams of longitudinal dimension much greater than the characteristic plasmon wavelength, the excitation of Langmuir wakes is negligible [103, 115], making it possible to assume that $\psi \equiv 0$. Under this assumption, (6.8) becomes $\varphi \equiv \gamma = \sqrt{1+|a|^2}$. Combining (6.7) and (6.10), one obtains the expression for the nonlinear term in (6.6) of the form

$$\frac{n}{\gamma} = \frac{1 + (\Delta_\perp + \partial_\eta^2)\gamma}{\gamma} \,. \tag{6.16}$$

Equations (6.6) and (6.16) make a closed nonlinear model of the interactions between a long relativistically intense laser beam and a cold underdense plasma. This model was investigated in [103, 115]. Its Lagrangian density is

$$L = \mathrm{i}a^* a_\tau - \varepsilon^2 |a_\eta|^2 - \frac{\varepsilon\theta}{2}(a_\eta a_\tau^* + c.c.) - (\nabla_\perp a, \nabla_\perp a^*) + |a|^2 + |\nabla_\perp \gamma|^2 + (\gamma_\eta)^2 - 2\gamma + \frac{\varepsilon^2}{4}|a_\tau|^2 \,, \tag{6.17}$$

and the corresponding invariants are the field mass given by (6.13), the momentum expressed as

$$\boldsymbol{P}_\perp = \int \Big[\mathrm{i}a^* \nabla_\perp a - \frac{\varepsilon\theta}{2}(a_\eta^* \nabla_\perp a + c.c.) + \frac{\varepsilon^2}{4}(a_\tau^* \nabla_\perp a + c.c.) \Big] \mathrm{d}^2 x_\perp \mathrm{d}\eta \,, \tag{6.18}$$

$$P_\parallel = \int \left[(\mathrm{i}a^* a_\eta - \varepsilon\theta |a_\eta|^2 + \frac{\varepsilon^2}{4}(a_\tau^* a_\eta + c.c.) \right] \mathrm{d}^2 x_\perp \mathrm{d}\eta \,,$$

and the Hamiltonian given by

$$H = \int \Big[\varepsilon^2 |a_\eta|^2 + (\nabla_\perp a, \nabla_\perp a^*) - |\nabla_\perp \gamma|^2 - (\gamma)_\eta^2 - |a|^2 + 2(\gamma - 1) + \frac{\varepsilon^2}{4}|a_\tau|^2 \Big] \mathrm{d}^2 x_\perp \mathrm{d}\eta \,. \tag{6.19}$$

The above model describes the scattering of laser radiation by the inhomogeneities of the plasma's dielectric response emerging due to the relativistic nonlinearity and ponderomotive charge displacement under conditions in which no significant generation of plasma waves occurs.

6.3 Envelope Approximation

Neglecting the terms responsible for high-order dispersion and the inertia of the laser-driven plasma electron component (since these terms are proportional to a small parameter) makes it possible to simplify the model embodied in (6.6)–(6.10) substantially.

The quantity γ^2 can be excluded from (6.6)–(6.10); the result is

$$\psi_\eta - \frac{1+|a|^2+|\nabla_\perp\psi|^2-\varphi^2}{2\theta\varphi} = \frac{\varepsilon^2\psi_\eta^2-(\varepsilon/2)^2\psi_\tau^2+\varepsilon\psi_\tau(\varphi+\theta\psi_\eta)}{2\theta\varphi}. \tag{6.20}$$

Substituting the expression for ψ_η, which follows from (6.8), in (6.9), we find that

$$\left(\frac{n}{\gamma}\varphi\right)_\eta - \theta(\nabla_\perp, \frac{n}{\gamma}\nabla_\perp\psi) = \frac{\varepsilon}{2}\left[\theta n_\tau + \left(\frac{n}{\gamma}\psi_\tau\right)_\eta\right] + \varepsilon^2 n_\eta\,. \tag{6.21}$$

Dropping the small terms in the right-hand sides of (6.6), (6.20), and (6.21), which corresponds to using the quasistatic approximation in fluid dynamics (for example, see [84]), and using the fact that, to the same order in ε, the group velocity is equal to the speed of light, $\theta \approx 1$, we establish the following set of nonlinear equations to describe the interactions of relativistically intense, ultrashort laser pulses with cold underdense plasmas in the framework of the envelope approximation [120]:

$$\mathrm{i}a_\tau + \triangle_\perp a + (1-F)a + \varepsilon^2 a_{\eta\eta} + \varepsilon a_{\eta\tau} = 0\,, \tag{6.22}$$

$$\psi_\eta = \frac{1+|a|^2+|\nabla_\perp\psi|^2-\varphi^2}{2\varphi}\,, \tag{6.23}$$

$$(\triangle_\perp + \partial_\eta^2)\varphi = \frac{F(1+|a|^2+|\nabla_\perp\psi|^2+\varphi^2)}{2\varphi} - 1\,, \tag{6.24}$$

$$(F\varphi)_\eta = (\nabla_\perp, F\nabla_\perp\psi)\,. \tag{6.25}$$

Here, we use the notation

$$F = \frac{n}{\gamma}\,. \tag{6.26}$$

The quantity F defined by the above equation is the normalized local value of the plasma frequency squared (the plasma frequency is modified by propagating laser radiation due to the relativistic increase in the electron mass and to charge-displacement). Following [119], we treat F as one of the variables in the framework of the model comprising (6.22)–(6.26).

Model (6.22)–(6.26) describes the interactions of ultrashort, relativistically intense laser pulses and cold underdense plasmas in the envelope approximation under the assumption that the inertia of the plasma electron component driven by the superintense electromagnetic field is negligible (quasi-static approximation). The powerful laser–plasma interaction phenomena included in the present model are relativistic nonlinearity, ponderomotive charge-displacement, generation of plasma waves by propagating laser pulse, as well as laser radiation diffraction and refraction.

The Lagrangian density for problem (6.22)–(6.26) is [120]

$$\begin{aligned} L = &\mathrm{i}a^* a_\tau - \varepsilon^2 |a_\eta|^2 - \frac{\varepsilon}{2}(a_\eta a_\tau^* + c.c.) - (\nabla_\perp a, \nabla_\perp a^*) \\ &+ |a|^2 + \varphi_\eta^2 + |\nabla_\perp \varphi|^2 \\ &+ 2\varphi(F\psi_\eta - 1) - F(1 + |a|^2 + |\nabla_\perp \psi|^2 - \varphi^2) \,. \end{aligned} \tag{6.27}$$

The corresponding invariants for (6.22)–(6.26) are the Hamiltonian of the model considered:

$$\begin{aligned} H = \int \Big[&\varepsilon^2 |a_\eta|^2 + (\nabla_\perp a, \nabla_\perp a^*) - \varphi_\eta^2 - |\nabla_\perp \varphi|^2 \\ &- |a|^2 - 2(\varphi - 1) \Big] \mathrm{d}^2 x_\perp \mathrm{d}\eta \,, \end{aligned} \tag{6.28}$$

the field mass

$$N = \int \left[|a|^2 + \frac{\mathrm{i}\varepsilon}{2}(a_\eta^* a - c.c.) \right] \mathrm{d}^2 x_\perp \mathrm{d}\eta \,,$$

and the momentum

$$\boldsymbol{P}_\perp = \int \left[\mathrm{i}a^* \nabla_\perp a - \frac{\varepsilon}{2}(a_\eta^* \nabla_\perp a + c.c.) \right] \mathrm{d}^2 x_\perp \mathrm{d}\eta \,,$$

$$P_\parallel = \int \left(\mathrm{i}a^* a_\eta - \varepsilon |a_\eta|^2 \right) \mathrm{d}^2 x_\perp \mathrm{d}\eta \,.$$

The above expression for the Hamiltonian shows that the model considered describes the interactions of positive and negative energy waves. The first and second terms of the Hamiltonian correspond to the laser radiation kinetic energy and the third and the fourth terms stand for the plasma wave kinetic energy; the plasma wave kinetic energy obviously is negative.

6.4 Long Beam and Large Aperture Limits

The general models considered allow substantial simplifications in two physically important cases, namely, when the laser pulse aperture or length is much greater than the characteristic plasma wavelength. These two limits are considered below.

6.4.1 Long Beam Limit

A large number of works addressed the problem of propagation of laser beams that are much longer than the plasmon characteristic wavelength (for example, see [7,95]). In this case, the major role is played by the effects related to the nonlinear evolution of the beam cross section, whereas the excitation of longitudinal plasma waves can be neglected. The corresponding set of equations can be derived from the model represented by (6.22)–(6.26). In the long pulse limit, $\partial_\eta = 0$, and there is no mechanisms of plasma wave generation, so that $\psi \equiv 0$. Then, from (6.23),

$$\varphi = \sqrt{1 + |a|^2}\,,$$

and, using (6.22) and (6.24), we establish the following set of self-consistent nonlinear equations:

$$\mathrm{i}a_\tau + \triangle_\perp a + (1 - F)a = 0\,, \tag{6.29}$$

$$F = \frac{\max\{0, 1 + \triangle_\perp \sqrt{1 + |a|^2}\}}{\sqrt{1 + |a|^2}}\,. \tag{6.30}$$

The operator $\max\{0, \cdot\}$ is introduced in (6.30) to guarantee that the plasma electron concentration, calculated with the help of this expression, is nonnegative. It was demonstrated in numerical simulations that electronic cavitation, i.e., the complete expulsion of plasma electrons from certain spatial domains, can occur due to the impact of the ponderomotive force in plasmas irradiated by intense laser pulses [7, 95]. Specifically, when the power of a propagating laser pulse is greater than the critical power of relativistic and charge-displacement self-focusing, cavitation occurs as the first focus emerges, and the laser radiation propagates further in a plasma channel. The immediate consequence of this phenomenon is that the nonlinearity is thus "switched off," which has an overall stabilizing influence on the propagation character.

The Lagrangian density for (6.29)–(6.30) is

$$L = \mathrm{i}a^* a_\tau - (\nabla_\perp a, \nabla_\perp a^*) + |a|^2 - 2\sqrt{1 + |a|^2} + \left|\nabla_\perp \sqrt{1 + |a|^2}\right|^2\,,$$

and the corresponding conserved quantities (Hamiltonian, field mass, and momentum) are expressed as [81,82,120]

$$H = \int \Big[(\nabla_\perp a, \nabla_\perp a^*) + 2(\sqrt{1 + |a|^2} - 1) - \left|\nabla_\perp \sqrt{1 + |a|^2}\right|^2\Big]\mathrm{d}^2 x_\perp\,, \tag{6.31}$$

$$N = \int |a|^2 \mathrm{d}^2 x_\perp\,, \tag{6.32}$$

$$\boldsymbol{P}_\perp = \mathrm{i}\int a^* \nabla_\perp a \mathrm{d}^2 x_\perp \, .$$

Problem (6.29)–(6.30) has been studied thoroughly. A countable set of eigenmodes of this problem is presented in [7, 95]. Another finding of these papers is the theoretical prediction of the relativistic and charge-displacement self-channeling of superintense, ultrashort, laser pulses in cold underdense plasmas. The essence of this phenomenon is that at large propagation distances, the transverse laser intensity distributions become independent of the depth of their penetration into plasmas and are described by the eigenmodes of problem (6.29)–(6.30). As demonstrated in numerical simulations, a threshold power exists for relativistic and charge-displacement self-channeling. Besides, in accord with [82], the sufficient condition for relativistic and charge-displacement self-channeling is $H < 0$. These issues will be discussed in finer detail in Chap. 8.

6.4.2 Large Laser Pulse Aperture Limit

In the large laser pulse aperture limit, $\nabla_\perp = 0$, and it follows from (6.25) that the variable F can be excluded:

$$F = \varphi^{-1} \, .$$

Under the same condition, (6.22) and (6.25) become

$$\mathrm{i}a_\tau + (1 - \varphi^{-1})a + \varepsilon^2 a_{\eta\eta} + \varepsilon a_{\eta\tau} = 0 \, , \tag{6.33}$$

$$\varphi_{\eta\eta} = \frac{1}{2}\left(\frac{1 + |a|^2}{\varphi^2} - 1\right) . \tag{6.34}$$

This model was presented and treated in [106]. Its Lagrangian density is

$$L = \mathrm{i}a^* a_\tau - \varepsilon^2 |a_\eta|^2 - \frac{\varepsilon}{2}(a_\eta a_\tau^* + \mathrm{c.c.}) + |a|^2 + \varphi_\eta^2 - \left(\frac{1 + |a|^2}{\varphi} + \varphi\right) ,$$

and the corresponding invariants (Hamiltonian, field mass, and momentum) are expressed as

$$H = \int \left[\varepsilon^2 |a_\eta|^2 - \varphi_\eta^2 - |a|^2 + \left(\frac{1 + |a|^2}{\varphi} + \varphi\right) - 2\right] \mathrm{d}\eta \, ,$$

$$N = \int \left[|a|^2 + \frac{\mathrm{i}\varepsilon}{2}(a_\eta^* a - \mathrm{c.c.})\right] \mathrm{d}\eta \, ,$$

$$P_\parallel = \int \left(\mathrm{i}a^* a_\eta - \varepsilon |a_\eta|^2\right) \mathrm{d}\eta \, .$$

If the laser pulse length is of unit order (of the order of the plasma's collisionless skin depth in terms of unnormalized variables) and the amplitude distribution does not exhibit rapid oscillations in η, the last two terms in

(6.33) can be dropped. In this case, the latter equation is easily integrated, so that

$$a = a_0(\eta) \exp\left(\mathrm{i}\left[1 - \varphi_0^{-1}(\eta)\right]\tau\right),$$

where $\varphi_0(\eta)$ is the solution of the following ordinary differential equation:

$$\varphi_{0_{\eta\eta}} = \frac{1}{2}\left(\frac{1 + |a_0|^2}{\varphi_0^2} - 1\right), \tag{6.35}$$

and $a_0(\eta)$ is an arbitrary function (determined by boundary or initial conditions). Note that in the particular case examined, the local value of the propagating laser pulse frequency shift is proportional to the local value of the plasma frequency squared. This fact is potentially useful for diagnostics of laser–plasma interactions at relativistic intensities. The plasma response equation (6.35) can be rewritten as

$$(\varphi_{0_\eta}^2 + \varphi_0 + \varphi_0^{-1})_\eta = \frac{|a_0|^2}{\varphi_0^2}\varphi_{0_\eta},$$

and thus we obtain the law of laser radiation energy transfer to plasma waves.

6.5 Filamentation and Self-Modulation of Relativistically Intense Laser Radiation in Cold Underdense Plasmas

In this section, the models of laser–plasma interactions at relativistic intensities developed above are applied to study the linear instabilities of superintense plane electromagnetic waves in a cold underdense plasma. As a preliminary study, this instability was treated in [118] assuming that the plasma electrons "follow" the electromagnetic field instantly and that no Langmuir waves are excited in plasma. The growth rate of instability was calculated with the help of a model based on a nonlinear wave equation with relativistic and charge-displacement nonlinearity (6.16). The growth rate of plane electromagnetic wave instability was calculated numerically in [121] in the framework of the model including relativistic and charge-displacement nonlinearities, generation of Langmuir waves in plasmas, high-order dispersion, and plasma electron component inertia. The corresponding instability describes the Raman scattering of laser radiation by plasmons and its Compton scattering.

An analytical expression can be derived for the growth rate of the instability of plane relativistically intense electromagnetic waves in plasmas in the slow field amplitude approximation. This calculation is presented below.

The exact plane wave solution of (6.22)–(6.26) is

$$a = a_0 \exp\left[\mathrm{i}(1 - \gamma_0^{-1})\tau\right],$$

$$\gamma_0 = \sqrt{1 + a_0^2}, \qquad \varphi = \gamma_0, \qquad \psi = 0, \qquad F = \gamma_0^{-1}$$

[one can verify easily that this solution differs from that given by (2.65)–(2.66) by terms of the order of ε, which is the precision of the Schroedinger equation approximation]. To study its instability we present the solution of (6.22)–(6.26) as a superposition of the above ground state and small perturbations

$$a = (a_0 + \delta a)\exp\left[\mathrm{i}(1-\gamma_0^{-1})\tau\right],$$
$$\varphi = \gamma_0, +\delta\varphi, \qquad \psi = \delta\psi, \qquad F = \gamma_0^{-1} + \delta F.$$

These perturbations obey the following linearized equations

$$\mathrm{i}\delta a_\tau + \triangle_\perp \delta a - a_0\delta F + \mathrm{i}(1-\gamma_0^{-1})\varepsilon\delta a_\xi + \varepsilon^2\delta a_{\eta\eta} + \varepsilon\delta a_{\eta\tau} = 0,$$

$$\delta\psi_\eta = \frac{a_0}{2\gamma_0}(\delta a + \delta a^*) - \delta\varphi,$$

$$(\triangle_\perp + \partial_\eta^2)\delta\varphi = \frac{a_0}{2\gamma_0^2}(\delta a + \delta a^*) + \gamma_0\delta F,$$

$$\gamma_0^{-1}\triangle_\perp\delta\psi = \gamma\delta F_\eta + \gamma_0^{-1}\delta\varphi_\eta.$$

Seeking their solutions in the form

$$\delta(a,\varphi,\psi,F)^T = \delta(a^0,\varphi^0,\psi^0,F^0)^T\exp\{\mathrm{i}\left[(\boldsymbol{k},\boldsymbol{x}_\perp) + \chi\eta - \Omega\tau)\right]\},$$

we arrive at the following expression for the eigenvalue Ω (the growth rate is given by $\operatorname{Im}\Omega$):

$$\Omega = \frac{\varepsilon\chi}{2\left[1-(\varepsilon\chi)^2\right]}\Big\{2\left[1-\gamma_0^{-1} - |\boldsymbol{k}|^2 - (\varepsilon\chi)^2\right] + R(a_0^2,|\boldsymbol{k}|^2,\chi^2)\Big\} \pm \sqrt{D(a_0^2,|\boldsymbol{k}|^2,\chi^2)}, \tag{6.36}$$

$$D = D_0 + (\varepsilon\chi)^2 D_1 + (\varepsilon\chi)^4 D_2, \tag{6.37}$$

$$D_0 = 4|\boldsymbol{k}|^2\left[|\boldsymbol{k}|^2 - R(a_0^2,|\boldsymbol{k}|^2,\chi^2)\right], \tag{6.38}$$

$$D_1 = \frac{8|\boldsymbol{k}|^2}{\gamma_0} + \left[R(a_0^2,|\boldsymbol{k}|^2,\chi^2) - \frac{4}{\gamma_0}\right], \tag{6.39}$$

$$D_2 = \frac{4}{\gamma_0^2}, \tag{6.40}$$

$$R(a_0^2,|\boldsymbol{k}|^2,\chi^2) = \frac{a_0^2(\gamma_0^{-1} + |\boldsymbol{k}|^2)}{\gamma_0^2(1-\gamma_0\chi^2)}. \tag{6.41}$$

The growth rate of laser radiation filamentation and self-modulation instability $\operatorname{Im}\Omega$ for $a_0 = 1$ and $\varepsilon = 0.2$ is shown in Fig. 6.1 as a function of k_1 and χ for $k_2 = 0$. Below, this growth rate is considered in some of the possible limiting cases.

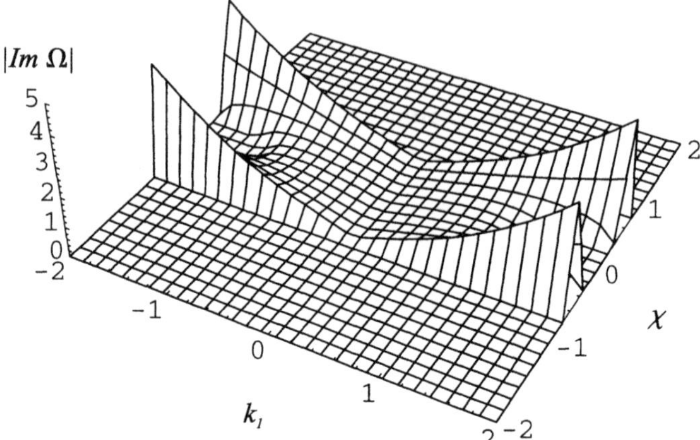

Fig. 6.1. The growth rate of laser radiation filamentation and self-modulation instability calculated with the help of (6.36)–(6.41) for $a_0 = 1$ and $\varepsilon = 0.2$ as a function of k_1 and χ for $k_2 = 0$ [120]

Long Perturbation Limit. In this limit, $\chi = 0$, i.e., there is no dependence on the longitudinal coordinate in the perturbations. This is the case of intense laser radiation filamentation instability; its growth rate is

$$\Omega_\perp = \pm \frac{|\boldsymbol{k}|}{\gamma_0} \sqrt{|\boldsymbol{k}|^2 - \frac{a_0^2}{\gamma_0}} \, .$$

The instability occurs when the following condition is met:

$$|\boldsymbol{k}|^2 < \frac{a_0^2}{\gamma_0} \, .$$

This result was obtained in [97] with the help of (6.29)–(6.30) (also see [120]). The above instability leads to the disintegration of the laser beam transverse cross section and the emergence of several light filaments. This instability picture is qualitatively similar to that described in [17] for cubic media. The power trapped in one filament estimated on the qualitative level on the basis of the above growth rate expression is [97]

$$P_{\text{filament}} = \frac{m^2 c^5}{e^2} \cdot \frac{\omega}{\omega_\text{p}} \sqrt{1 + I/I_\text{r}} \, ,$$

(in dimensional form) where I is the electromagnetic wave intensity, $I_\text{r} = m^2\omega^2 c^3/(4\pi e^2)$ is the relativistic intensity, and e is the electron charge.

Large Perturbation Aperture Limit. In this case, $\boldsymbol{k} = 0$, i.e. the perturbations are independent of the transverse coordinates. This limit pertains to self-modulation instability.

First, we consider the case where the perturbation characteristic scale is of the same order as the plasma's collisionless skin depth, namely, $\chi = O(1)$. It follows easily from (6.36)–(6.41) that

$$\Omega_{\parallel} = \frac{\varepsilon\chi\{2\left[1-\gamma_0^{-1}-(\varepsilon\chi)^2\right] + R(a_0^2,0,\chi^2)\} \pm \sqrt{D_{\parallel}(a_0^2,\chi^2)}}{2\left[1-(\varepsilon\chi)^2\right]},$$

$$D_{\parallel}(a_0^2,\chi^2) = R(a_0^2,0,\chi^2)\left[R(a_0^2,0,\chi^2) - \frac{4}{\gamma_0}\right] + \left(\frac{2\varepsilon\chi}{\gamma_0}\right)^2,$$

and, therefore $\Omega_{\parallel} = O(\varepsilon)$. One can conclude that for large aperture perturbations, the growth rate of self-modulation instability tends to be small. Now, we consider the case where $\varepsilon\chi = O(1)$. Then, $R(a_0^2,0,\chi^2) = O(\varepsilon^2)$, and $D_{\parallel} = O(\varepsilon)$, so that again the magnitude of the growth rate is of the order of the small parameter.

Combined Filamentation and Self-Modulation Instability. The most interesting case of the instability of plane relativistically intense electromagnetic fields in cold underdense plasmas is the combination of filamentation and self-modulation instabilities. Assume again that $\chi = O(1)$. It follows from the general expression for the growth rate that

$$\Omega = \pm|\boldsymbol{k}|\sqrt{|\boldsymbol{k}|^2 - R(a_0^2,|\boldsymbol{k}|^2,\chi^2)} + O(\varepsilon).$$

The instability conditions are

$$0 < \gamma_0\chi^2 < 1$$

for $|\boldsymbol{k}|^2 < a_0^2/\gamma_0$, and

$$\gamma_0^{-2}\left(1 - \frac{a_0^2}{\gamma_0|\boldsymbol{k}|^2}\right) < \gamma_0\chi^2 < 1$$

for $|\boldsymbol{k}|^2 > a_0^2/\gamma_0$.

Note that the first of the above two inequalities is associated with the values of the perturbation wave vector transverse components $\boldsymbol{k}$ for which the relativistically intense electromagnetic field in plasma is unstable in the purely transverse case, whereas the second involves the values of the transverse wave vector components for which plane intense electromagnetic waves are stable in plasma when there is no dependence on the perturbation wave vector's longitudinal component. Obviously, in three-dimensional geometry, first, filamentation instability is neutralized for $\chi^2 < \chi^2_{\text{resonant}} = \gamma_0^{-1}$, and, second, an additional electromagnetic field instability mechanism emerges.

Note that it follows from the above instability conditions that plane intense electromagnetic waves in plasmas are stable for $\chi^2 > \gamma_0^{-1}$. Since $\gamma_0^{-1} \to 0$ for $a_0 \to \infty$, extremely intense electromagnetic waves in cold underdense plasmas tend to be stable. This fact is explained by the nonlinearity "shutdown" when the electrons become extremely heavy in such fields.

6.6 Laser Radiation Stimulated Scattering by Plasmons and Third-Harmonics Generation

A laser radiation instability study similar to that described above was performed in [121] on the basis of the model represented by (6.6)–(6.10). The growth rate calculated in this paper is depicted in Fig. 6.2 for $k_2 = 0$. The notations in this figure are $k_1' = k_1/k_0$ and $\chi' = \chi/k_0$. The growth rate distribution has a figure-eight shape oriented along the χ axis.

The scattered radiation' angular distribution can be calculated from the growth rate. In the laboratory reference frame, the scattered wave phase can be written as

$$\Phi_1 = \Big[(\boldsymbol{k}_\perp, \boldsymbol{x}_\perp) + (\chi + k_0)x_3 - (\chi\theta + \omega_0 - \varepsilon\Gamma/2 \\ + \max \mathrm{Im}\lambda\varepsilon/2)t + \max \mathrm{Re}\lambda(\varepsilon/2)t\Big] ,$$

where λ is the stability problem eigenvalue and

$$\Gamma = \frac{2}{\varepsilon^2}\left(1 - \sqrt{1 - \varepsilon^2(1 - \gamma_0)}\right) .$$

The scattered wave propagates at an angle to the x_3 axis which is given by $\tan\phi = k_\perp/(k_0 + \chi)$. For lower radiation intensities, the growth rate distribution retains its figure-eight shape, but the spread of the rings decreases.

Figure 6.3 shows the cross section of the growth rate distribution along the χ axis. In this case, the laser–plasma interaction model (6.6)–(6.10) involves no additional assumptions since exact conservation of circular polarization is possible in one-dimensional geometry.

The results presented in Fig. 6.3 agree exactly with those of [61] which were obtained directly from Maxwell and fluid dynamics equations. There are

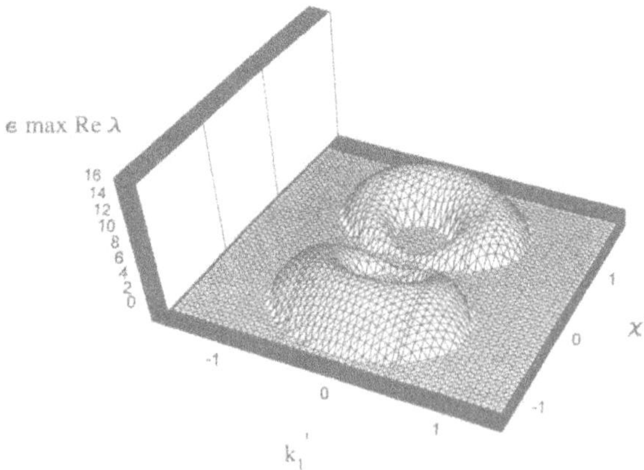

Fig. 6.2. Distribution of the growth rate $\varepsilon \max \mathrm{Re}\,\lambda$ calculated on the basis of problem (6.6)–(6.10) for $a_0 = 1$ and $\varepsilon^2 = 7.4 \times 10^{-2}$ [121]

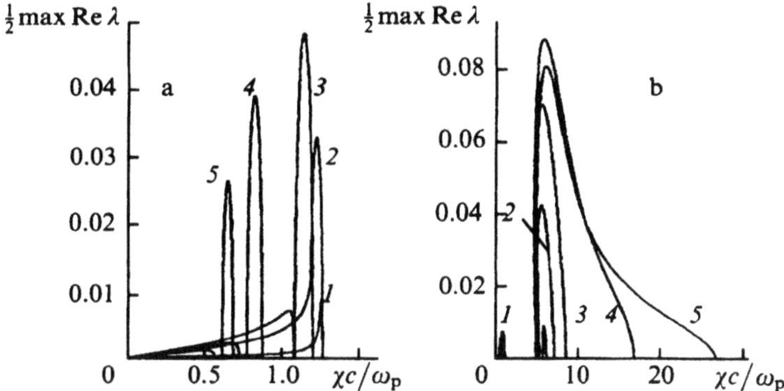

Fig. 6.3. Growth rate distribution cross section along the χ-axis for $\varepsilon^2 = 10^{-2}$ (**a**) and $\varepsilon^2 = 7.4 \times 10^{-2}$ (**b**) corresponding to the following ground state amplitudes: (1) $a_0 = 0.1$, (2) $a_0 = 0.5$, (3) $a_0 = 1$, (4) $a_0 = 3$, and (5) $a_0 = 5$ [121]

four local maxima in the growth rate distribution presented in Fig. 6.3; two of them correspond to positive values of χ and two to negative values. Let us number them by an index i in consecutive order from left to right. The wave vector of the scattered wave $\boldsymbol{k}'$ is obtained by adding the perturbation wave vector to the initial wave vector $\boldsymbol{k} = \boldsymbol{e}_3 k_0$. As we see in Fig. 6.3, the $i = 1$ local maximum corresponds to $\boldsymbol{k}' = -\boldsymbol{k}$ which is the signature of backward scattering at the fundamental frequency. The $i = 2, 3$ local maxima are related to forward stimulated scattering with $\boldsymbol{k}' = \boldsymbol{k} \mp \boldsymbol{k}_\mathrm{p}$, and the last local maximum ($i = 4$) describes forward third-harmonics generation: $\boldsymbol{k}' \approx 2\boldsymbol{k}$ and $\omega' \approx 2\omega_0$. This interpretation is supported by the calculation of phase velocities: it is found that the phase velocity is negative for the first local maximum and positive for the others.

Figure 6.4(a) illustrates the maximum growth rate angular dependencies for the above four scattering branches. For forward and backward scattering angles, $\phi \approx 50°$ and $2\phi \approx 210°$, respectively.

A study performed with the model ignoring the plasma oscillation excitation [118] also showed forward third-harmonics generation and fundamental frequency backscattering. Consequently, these effects cannot be attributed to the emergence of plasma waves during laser pulse propagation but are due to the relativistic nonlinearity embodied in the mass factor γ. This nonlinearity introduces double-frequency oscillations in the linearized equations; the corresponding perturbations propagate forward and backward. The addition of the fundamental phase and the phases of these perturbations results in the emergence of third harmonics and of the backscattered wave.

It is informative to consider the above results in the context of the geometry of a real experiment. The local gain value is related to the instability growth rate. At every point in space in the inhomogeneous plasma, the gain is a function of the electromagnetic wave frequency and propagation direction.

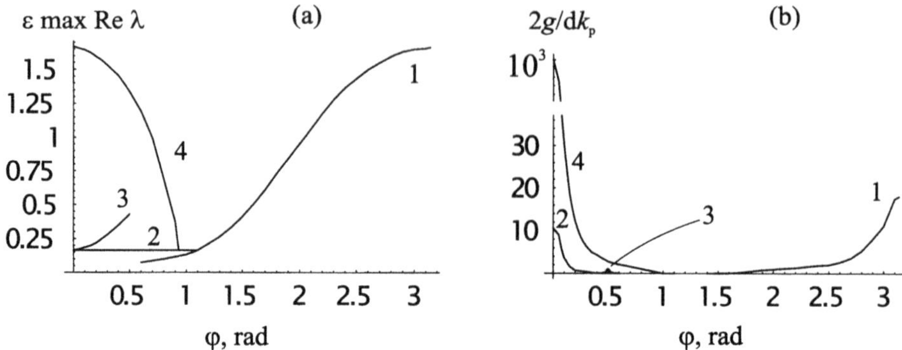

Fig. 6.4. (a) Maximal growth rate angular dependencies for scattering at the fundamental frequency (1); stimulated Raman scattering at $\omega \pm \omega_{\mathrm{p}}$ (2,3); and third-harmonics generation (4). **(b)** Geometrical optics calculation of the gain scattering angle dependence for a homogeneous cylindrical filament of a 1 cm length and a 4 µm diameter for a laser pulse with the longitudinal size of 40 µm [121]

At present, it is possible to form plasma filaments that span up to 1 cm and have a diameter of $d = (5-100)\,\lambda$ for laser pulses of lengths $L_{\mathrm{p}} < 3\times10^{-2}$ cm. The gain $g = kL(\phi)$ for the four scattering branches and for the above plasma filament parameters is depicted in Fig. 6.4(b).

In a filament, the amplification of forward scattered radiation occurs in a narrow cone with a vertex angle of $2\phi \approx d/L$. Stimulated Raman scattering and third-harmonics generation take place within this cone. The third harmonic is scattered into an angle of $2\phi = 60°$ without significant amplification. The laser radiation is scattered backward at the fundamental frequency into a wide solid angle, but here the amplification efficiency and the scattered radiation intensity tend to be low since the amplification distance is very short. Naturally, all of these considerations are purely qualitative in character because under real experimental conditions, radiation diffraction enhances scattering angles.

Another issue that deserves attention is the dependence of the scattering picture on the value of the parameter ε which stands for the ratio of unperturbed plasma frequency and that of laser radiation. The growth rate is maximal for $\varepsilon = 0.3333$ and exceeds that discussed above by approximately 25%. The growth rate dependence on ε is illustrated by Fig. 6.5.

The scattering mechanism can be interpreted as follows. The trajectories of electrons driven by a circularly polarized electromagnetic field are circles located in planes perpendicular to the propagation direction. These electrons emit an infinite spectrum of harmonics with frequencies $\omega_n' = n\omega - \Delta\omega$, where the frequency shift $\Delta\omega$ is induced by the electron response. The angular radiation distribution at the frequency ω_{n}' is anisotropic, but the radiation is emitted into a 4π solid angle. The forward and backward half-space angular distributions are symmetrical. Plasma oscillations at frequency ω_{p} are also

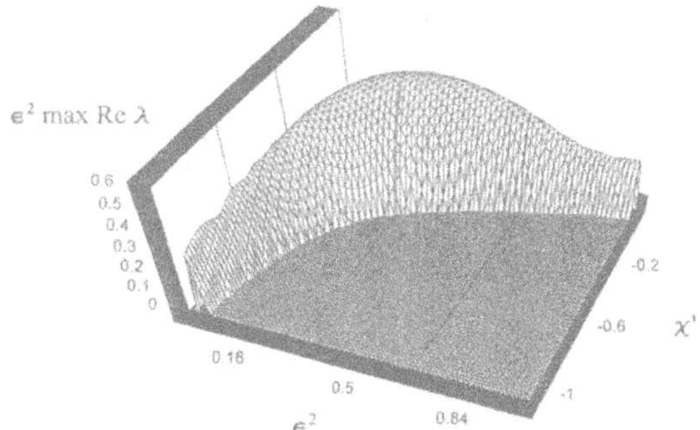

Fig. 6.5. Growth rate distribution as a function of ε^2 and χ for $\boldsymbol{k}_\perp = 0$ at $a_0 = 1$. $\chi' = 10.61 \times \chi$

present in the medium. In a cold plasma, oscillations are standing waves of arbitrary spatial wavelengths. These standing waves can be presented as superpositions of pairs of counterpropagating waves. This system supports a range of wave processes. First of all, it is the second harmonics ($\omega + \omega = 2\omega$) and slow electromagnetic field ($\omega - \omega = 0$) generation, as well as the third-harmonics generation and backscattering ($\omega - 2\omega = -\omega$). These processes are characterized by rigorous phase matching conditions, but the phase matching condition regions are broadened due to fluid dynamics and thermal oscillations. In the problem considered, these interactions result in the emergence of plasma waves in accord with the following scheme: $\omega + \omega \pm \omega_\mathrm{p} = 2\omega \pm \omega_\mathrm{p}$, $\omega - \omega \pm \omega_\mathrm{p} = \omega_\mathrm{p}$, $\omega + 2\omega \pm \omega_\mathrm{p} = 3\omega \pm \omega_\mathrm{p}$, $\omega - 2\omega \pm \omega_\mathrm{p} = -\omega \pm \omega_\mathrm{p}$, etc.

6.7 Conclusions

A hierarchy of models describing the time-dependent three-dimensional nonlinear propagation of relativistically intense, ultrashort laser pulses in cold underdense plasmas is presented in this chapter. The basic model accounts for the following phenomena: laser radiation diffraction and refraction, relativistic and charge-displacement nonlinearities resulting from modification of the plasma's dielectric response due to the increase in the masses of free electrons driven by intense electromagnetic fields and to the impact of the ponderomotive force on the spatial profile of the electron concentration, and generation of waves in the plasma's electron component. Lagrangians and invariants of the models are presented. These models are applied to study a number of types of relativistically intense laser radiation instabilities in cold underdense plasmas. As a concluding remark, note that although the advantage of studying electromagnetic field instabilities in plasmas in the framework of powerful laser–plasma interaction models is the relative simplicity and ease of interpreting results, a more profound understanding of this range of phenomena can be attained by using Maxwell and fluid dynamics equations, which was done in Chaps. 4 and 5. The most significant applications of the models developed in this chapter are in simulations of nonlinear propagation of relativistically intense laser pulses in cold underdense plasmas.

7. Intense Laser Pulse Solitons in Plasmas

In the previous chapter, we considered intense laser–plasma interactions under the conditions in which the major role is played by the effects related to laser beam transverse aperture evolution. Now we turn to the opposite limit, where the laser pulse aperture is large compared to the characteristic plasma wavelength. A number of studies are dedicated to the interactions of large aperture intense laser pulses with cold underdense plasmas. In particular, energy transfer from the laser pulse to plasma waves is studied in [106] in the framework of model (6.33)–(6.34).

A remarkable class of slab geometry solutions of the problem of laser–plasma interactions at relativistic intensities is developed in [45]. Relativistically intense one-dimensional solitons with discrete amplitudes coupled to the waves in the plasma electron fluid are presented in that paper. Later the results of [45] were confirmed in [46]. Potentially, such solitons are useful as the basis for photon accelerator schemes.

7.1 Soliton Equations and Numerical Solutions

Consider the Maxwell equations and the equations of fluid dynamics (2.56)–(2.60) in one-dimensional geometry. Let us restrict ourselves to circular polarization and seek solutions of the above set of equations in the form (6.1), where the electromagnetic field amplitude, the scalar potential, the plasma electron concentration and momentum are functions of just one variable η defined by (6.2). The group velocity is $\theta = k/\omega$. In this case, the form of the nonlinear ordinary differential equations for the above functions and their derivation are largely similar to those of the Akhiezer–Polovin problem described in detail in Chap. 3. Specifically, for the vector potential amplitude and the scalar potential, we have the following coupled equations:

$$(1-\theta^2)U_{\eta\eta} + \left[s^2 - \frac{\theta(1-\theta v)}{(\theta - v)\varphi}\right] U = 0\,, \tag{7.1}$$

$$\varphi_{\eta\eta} = \frac{v}{\theta - v}\,, \tag{7.2}$$

where v is the plasma electron longitudinal velocity, which is expressed in terms of the vector and scalar potentials as

$$v = \frac{\theta(1+U^2) - \varphi\sqrt{\varphi^2 + (1-\theta^2)(1+U^2)}}{\varphi^2 + \theta^2(1+U^2)}, \tag{7.3}$$

and the real-valued parameter s^2 is given by

$$s^2 = \omega^2(1-\theta^2). \tag{7.4}$$

Naturally, it must be assumed that the group velocity is less than the speed of light, namely, $\theta < 1$.

Note that ion motions are neglected in the derivation resulting in the above equations. The more general problem formulated in [45] includes the additional terms related to the perturbation of the plasma ion background. We follow a more concise study presented in [46], where the ions are assumed to be immobile due to the short laser pulse duration.

Numerical studies of solutions of (7.1)–(7.4) which decay exponentially at $\eta \to \infty$ are described in [45, 46]. The major result established in these papers is that at sufficiently high laser pulse amplitudes, this problem is an eigenvalue one; for any particular value of θ, spatially localized solutions exist for discrete values of the parameter s^2 (the low amplitude limit is discussed in the following section). To follow the notations of [46], let us introduce another eigenvalue parameter $\lambda = k(q-\theta)$, where $q = \omega/k$ is the laser radiation's phase velocity. This parameter is related to s^2 as $s^2 = \omega\lambda$. The shapes of intense laser pulse solitons depend on the value of λ, but in typical examples presented in [45,46], the φ-distributions are bell-shaped, whereas the function U has numerous nodes. A soliton for $\theta = 0.97$ (for an underdense plasma the regime of interest is that where the group velocity is close to the speed of light) and $\lambda = 0.224445$ is depicted in Fig. 7.1 [46].

An important feature illustrated by Fig. 7.1 is the oscillatory character of the electromagnetic field amplitude. Identifying the problem parameters θ and q as the group and phase velocities, we used the corresponding definitions from linear theory. Nonlinear corrections to these quantities arise due to the

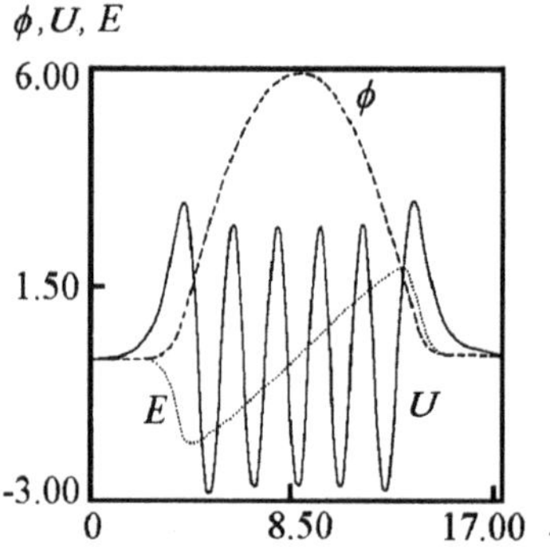

Fig. 7.1. Intense one-dimensional laser pulse soliton in a cold underdense plasma. The scalar potential, vector potential, and the electric field are shown [46]

above "additional" electromagnetic field oscillations which result when the laser pulse couples with the plasma waves. The corresponding nonlinear group velocity is closer to the speed of light than the linear one since both relativistic and charge-displacement nonlinearities tend to weaken the plasma's dielectric properties. According to the quantitative estimates of [46], where this issue is considered in depth, the deviations between the linear and nonlinear group velocities make up to 25% of the difference between these values and the speed of light.

7.2 One-Dimensional Laser Pulse Solitons in the WKB Approximation

Since, in the case of interest, the laser pulse central frequency is much greater than the speed of light, the above soliton problem can be studied in the framework of the spatially one-dimensional model embodied in (6.33)–(6.34), which are derived specifically in this limit. Let us develop solutions of these equations of the form

$$a(\eta,\tau) = U(\eta)\exp[\mathrm{i}(s-1)(\eta-2\tau)]\,, \qquad \varphi = \varphi(\eta)\,, \tag{7.5}$$

where U is real and s is a real-valued parameter. Substituting (7.5) in (6.33)–(6.34), we obtain

$$\epsilon^2 U_{\eta\eta} + P^2(\varphi)U = 0\,, \tag{7.6}$$

$$\varphi_{\eta\eta} = \frac{1}{2}\left(\frac{1+U^2}{\varphi^2} - 1\right), \tag{7.7}$$

with $P^2(\varphi) = s^2 - \varphi^{-1}$. The conservation law for the above coupled equations is written as

$$\varphi_\eta^2 - \epsilon^2 U_\eta^2 + V(U,\varphi) = E\,, \tag{7.8}$$

$$V(U,\varphi) = \varphi + \varphi^{-1} - U^2(s^2 - \varphi^{-1})\,. \tag{7.9}$$

We seek a bounded solution of (7.6) and (7.7) such that

$$\varphi(\infty) - 1 = a(\infty) = \varphi_\eta(\infty) = a_\eta(\infty) = 0\,. \tag{7.10}$$

These boundary conditions determine $E = 2$.

Low Amplitude Limit. For low laser pulse amplitudes, one can assume that the plasma remains quasi-neutral and set $\varphi_{\eta\eta} = 0$ in (7.7), which immediately gives $\varphi = \sqrt{1+U^2}$ [45]. Expanding this expression in U^2 and substituting the result in (7.6), one arrives at the following equation for the electromagnetic field amplitude:

$$\epsilon^2 U_{\eta\eta} + (s^2-1)U + \frac{1}{2}U^3 = 0\,. \tag{7.11}$$

It has the well-known soliton solution

$$U(\eta) = 2\sqrt{1-s^2}\frac{1}{\cosh\frac{\sqrt{1-s^2}\eta}{\epsilon}} . \tag{7.12}$$

WKB Approximation. Consider the more interesting case of high electromagnetic field amplitudes. Since θ is close to unity for substantially underdense plasmas, $\epsilon^2 = 1 - \theta^2 \ll 1$. We may solve (7.6) and (7.7) by taking advantage of the latter circumstance (the entire situation is similar to that considered in dealing with the Akhiezer–Polovin problem in Chap. 3). Equation (7.6) is isomorphic to the Schroedinger equation for a particle of momentum P and energy s^2 in a static potential well $1/\varphi$ [47]. Let φ have a maximum at $\eta = 0$, i.e., $\varphi(0) = \varphi_0$, $\varphi_\eta(0) = 0$, and let φ fall off monotonically to unity at $|\eta| \to \infty$. Boundary conditions (7.10) require exact symmetry about $\eta = 0$ for φ and U^2, i.e.,

$$\varphi(\eta) = \varphi(-\eta) , \qquad U(\eta) = \pm U(-\eta) , \tag{7.13}$$

and the turning points are determined by setting $P^2 = 0$ at $\eta = \pm\eta_0$. Because $\epsilon^2 \ll 1$, the WKB solution of (7.6) is

$$U(\eta) = \frac{g}{\sqrt{P(\varphi)}} \sin\left[\frac{1}{\epsilon}\int_0^\eta P(\varphi)\mathrm{d}\eta + \sigma\right] , \tag{7.14}$$

where g is a real constant, $\sigma = \pi/2$ for symmetrical solutions, and $\sigma = 0$ for the antisymmetric ones. Due to (7.13), we consider only the range $0 < \eta < \infty$. Substituting (7.14) in (7.8), we obtain to the lowest order in ϵ [47],

$$\varphi_\eta^2 = F(\varphi) = 2 - \varphi - \frac{1}{\varphi} + s^2\sqrt{s^2 - \frac{1}{\varphi}} , \tag{7.15}$$

$$g^2 = \frac{\varphi_0 + \varphi_0^{-1} - 2}{\sqrt{s^2 - \varphi_0^{-1}}} . \tag{7.16}$$

It can be established that the WKB solution within $\pm|\eta_0|$ is valid for [47]

$$\epsilon^2 \le \varphi_0 - 1 \ll \epsilon^{-2} . \tag{7.17}$$

We note that $F(\varphi)$ defined by (7.15) has two roots φ_0 and φ_1 with $1 < s^2\varphi_1 < s^2\varphi_0$, $\eta_1 < \eta_0$, $\eta_0 - \eta_1 \ll 1$. Thus φ_η^2 becomes negative within the range restricted by the WKB turning points. Special analysis is needed for the range $\eta \simeq \eta_1 < \infty$.

Equations (7.6)–(7.8) with $E = 2$ are equivalent to a single equation,

$$Z_{\psi\psi}^2 = \frac{4}{\epsilon^2\psi^4}\frac{Z_\psi[Z + (s^2 - \psi)Z_\psi]}{\psi + \psi^{-1} - Z - 2} , \tag{7.18}$$

for the function $Z(\psi) = \varphi_\eta^2 + \varphi + \varphi^{-1} - 2 = \epsilon^2 U_\eta^2 + (s^2 - \varphi^{-1})U^2$, where $\psi = \varphi^{-1}$. In this case, $U^2 = -Z_\psi$. Equation (7.18) must be solved with the boundary conditions $Z(0) = Z_\psi(0) = 0$, $Z(\psi_0) = Z_0 = (1 - \psi_0^2)/\psi_0$, and (i) $Z_\psi(\psi_0) = 0$ or (ii) $Z(\psi_0) = Z_0/(\psi_0 - s^2)$. Condition (i) leads to antisymmetrical solutions for $U(\eta)$, and (ii) to symmetrical solutions.

It can be shown that under condition (7.17), the parameter $s^2 = 1 - \Delta$ is close to unity, so that $\Delta \ll 1$, and in the vicinity of the turning point η_0, as well as in the whole "outer" range $\eta_0 < \eta < \infty$, φ is also close to unity. It is established in [47] (also, see references therein) by matching the solutions of (7.18) corresponding to the $\delta \ll \Delta$ and $\delta - \Delta > (\epsilon g)^{4/5}$ limits, that $\Delta = 1 - s^2 \approx 0.3\,(\epsilon g)^{4/5}$. By using a perturbation theory with respect to small Δ, one finds, to the zeroth order, that $g \approx [(\varphi_0 - 1)^3/\varphi_0]^{1/4}$ and

$$\Delta \approx 0.3\frac{\epsilon^{4/5}(\varphi_0 - 1)^{3/5}}{\varphi_0^{1/5}}\,, \qquad U_{\max} \approx 3.4\frac{\Delta}{\epsilon}\,, \tag{7.19}$$

where $U_{\max}$ is the absolute maximum of the vector potential envelope that is reached at the local maximum of $U(\eta)$ next to the WKB turning point (see [47] and the references therein).

By matching the phase from (7.14) with that for the $\delta - \Delta > (\epsilon g)^{4/5}$ region, one can obtain another relation corresponding to the WKB phase integral:

$$\int_{\varphi_1}^{\varphi_0} \sqrt{s^2 - \varphi^{-1}}\frac{\mathrm{d}\varphi}{\varphi_\eta} = \epsilon(N + k)\frac{\pi}{2}\,, \tag{7.20}$$

where the phase shift k differs slightly from the standard WKB value of 1/2: $k(N)$ varies slowly from $k_1 \approx 0.4$ for $N \leq 10$ to $k_2 \approx 0.2$ for large N.

The integral in (7.20) is insensitive to contributions from the region $\varphi \approx \varphi_1$. Evaluating it in two limiting cases $\varphi_0 - 1 \ll 1$ and $\varphi_0 \gg 1$ and interpolating between the two asymptotics, we obtain, to the zeroth order approximation,

$$\varphi_0 - 1 = K\epsilon^2(N + k)^2\,, \tag{7.21}$$

where $k(N)$ monotonically decreases from $K_1 \approx 1.11$ in the weakly nonlinear limit, $\epsilon^2 \simeq \varphi_0 - 1 \ll 1$ $(1 \leq N \ll \epsilon^{-1})$, to $K_2 \approx 0.62$ in the strongly nonlinear limit, $1 \ll \varphi_0 \ll \epsilon^{-2}$ $(\epsilon^{-1} \ll N \ll \epsilon^{-2})$. To the same accuracy,

$$\Delta = 1 - s^2 \approx \frac{0.3K^{3/5}\epsilon^2(N + k)^{6/5}}{[1 + K\epsilon^2(N + k)^2]^{1/5}}\,, \tag{7.22}$$

$$U_{\max} \approx \frac{K^{3/5}\epsilon^2(N + k)^{6/5}}{[1 + K\epsilon^2(N + k)^2]^{1/5}}\,. \tag{7.23}$$

Thus, in the range of φ_0 given by (7.17), the quantities φ_0, Δ, and $U_{\max}$ are quantized. The approximate relations (7.21) and (7.22) provide a *quantitative* basis for the discrete soliton spectrum in the general case of $1 \leq N \ll \epsilon^{-2}$ solitons.

The soliton width, $L \approx \int_{-\eta_1}^{\eta_1} \mathrm{d}\eta \approx 2\int_1^{\varphi_0} \mathrm{d}\varphi/\varphi_\eta \approx 4[(\varphi_0 - 1)^{1/2} + \pi/3]$ increases with amplitude unlike the conventional hyperbolic secant soliton

$N = 0$ discussed above. However, in the weakly nonlinear limit $\sqrt{\varphi_0 - 1} \ll 1$, the width is approximately constant, $L \approx 4\pi/3$. This is the absolute minimum of the soliton width (in dimensionless coordinate η).

Thus, by using asymptotic analysis with respect to the small parameter ϵ, one-dimensional soliton solutions are found analytically for intense laser pulses propagating in a uniform plasma. For large amplitude solitons, two nonlinear dispersion relations, (7.19) and (7.20) result in the discrete spectra given by (7.21) and (7.22).

7.3 Conclusions

A general analytical framework is developed in [47] for the nonlinear dispersion relations of a class of large amplitude, one-dimensional, isolated envelope solitons for modulated light pulses coupled to electron plasma waves, previously investigated numerically [45, 46]. The analytical treatment of weakly nonlinear solitons [142] is extended to the strongly nonlinear limit.

The principal features of these solitons are as follows:

1. The solitons are classified by the integer N which is the number of nodes of the envelope of the laser vector potential. An eigenvalue, $0 < s^2 \equiv 1 - \Delta < 1$, related to the nonlinear phase shift, is associated with each soliton.
2. As the laser vector potential amplitude is increased, the spectra of eigenvalues Δ and soliton amplitudes φ_0 are discrete. The first antisymmetric soliton with $N = 1$ occurs when $\Delta = D_1\epsilon^2$, where $D_1 = 0.5$. Low N discrete levels were studied by Kuehl and Zhang in the limit $1 \ll N \ll \epsilon^{-1}$. The treatment discussed in [47] also covers strongly nonlinear solitons with $\epsilon^{-1} < N \ll \epsilon^{-2}$.
3. In the large amplitude limit, $\Delta \approx 0.25\epsilon^{8/5}N^{4/5}$. Unlike the $N = 0$ soliton, the width L of the $N \gg \epsilon^{-1}$ solitons increases with the soliton amplitude: $L \sim \sqrt{\varphi_0}$.

8. Relativistic and Charge-Displacement Self-Channeling of Intense Ultrashort Laser Pulses in Plasmas

Relativistic and charge-displacement self-channeling is one of the most significant phenomena in the entire area of the interactions of extremely intense laser pulses with matter. Both the relativistic increase in the mass of laser-driven plasma electrons and the reduction of electron concentration under the effect of the ponderomotive force tend to reduce the local electron frequency in a laser-irradiated plasma. As a result, the simultaneous modification of the plasma dielectric response by the propagating electromagnetic radiation via the above two mechanisms enhances the medium transparency in the paraxial domain where the intensity is maximal. Under certain conditions discussed in this chapter, this effect leads to the emergence of channeled regimes of superintense laser pulse propagation in underdense plasmas. In these regimes, the laser radiation diffraction is suppressed, nonlinear beam self-trapping occurs, and the laser pulse propagates into the plasma over dozens of Rayleigh ranges. An additional electromagnetic radiation confinement mechanism emerges due to electron cavitation, namely, the total expulsion of the electron fluid from a certain spatial area. Then, the laser beam propagates within the resulting channel, where the intensities can reach the level of $10^{22}\,\mathrm{W/cm^2}$.

In this chapter, we present a theory of the relativistic and charge-displacement self-channeling of superintense laser pulses in cold underdense plasmas, including analytical results and simulations in the framework of the nonlinear Schroedinger equation model developed in the previous chapter.

8.1 Stationary Self-Localized Modes of Beam Propagation

A large number of nonlinear optics studies performed on the basis of nonlinear Schroedinger equation models is dedicated to the phenomena described by its stationary solutions. In particular, the cubic Schroedinger equation describes slab geometry optical solitons [18, 21], and this equation admits a countable set of nonlinear axially symmetrical eigenmodes [18, 19]. Below, we shall see that the nonlinear Schroedinger equation model of relativistic and charge-displacement laser beam propagation in a cold underdense plasma (6.30)–(6.30) allows solutions of similar types and that some of these solutions are extremely useful in describing the nonstationary evolution of relativistically intense optical fields in matter as well.

Consider solutions of (6.29)–(6.30) of the form

$$a = U(\boldsymbol{x}_\perp - \boldsymbol{v}_0 t) \exp\left\{\mathrm{i}\left(-(\boldsymbol{v}_0, \boldsymbol{x}_\perp)/2 + (s + |\boldsymbol{v}_0|^2/4)t\right)\right\} .$$

Here, $\boldsymbol{v}_0$ is a constant vector with real-valued components, and s is a real-valued parameter. With this choice of phase, amplitude U is also a real-valued function; the equation for it is

$$\triangle_\perp U + [s + F(U^2)]U = 0 , \tag{8.1}$$

$$F = -\frac{n}{\gamma} , \tag{8.2}$$

$$n = \max\{0, 1 + \triangle_\perp \gamma\} , \tag{8.3}$$

$$\gamma = \sqrt{1 + U^2} . \tag{8.4}$$

In the above set of equations, n is the electron concentration perturbed by the laser beam ponderomotive force, and γ is the relativistic mass factor. As discussed in Chap. 6, the above nonlinearity results from the local balance between ponderomotive and electrostatic forces, and the function $\max\{\cdot, 0\}$ is introduced to rule out physically unacceptable negative values of n. Below, we will be interested in self-localized solutions of the above equation, namely, such solutions for which the power integral (6.32) is finite.

8.1.1 Slab Geometry Solitons

Following [89], let us consider the slab geometry case where the self-localized beam propagation mode amplitude U depends on one Cartesian variable $x_1 - v_0 t$. In Sect. 2.4.2, we already encountered the set of ordinary differential equations identical to that which follows from (8.1)–(8.4) in this case [see (2.71)–(2.73)]. Using the results outlined in this section, one finds easily that the self-localized slab geometry beam propagation eigenmodes are superintense optical solitons whose amplitudes are given by [89]

$$a_0(x_1 - v_0 t) = \frac{2\sqrt{1 - s^2} \cosh\left(\sqrt{1 - s^2}(x_1 - v_0 t)\right)}{\cosh^2\left[\sqrt{1 - s^2}(x_1 - v_0 t)\right] + s^2 - 1} .$$

Readers can find a treatment of other types of slab geometry solutions of (6.30)–(6.30) in [89].

8.1.2 Axially Symmetrical Eigenmodes: Relativistic and Charge-Displacement Self-Channeling Critical Power

Axially Symmetrical Eigenmodes. Let us investigate the axially symmetrical case of problem (8.1)–(8.4), namely, let us set $\boldsymbol{v}_0 = 0$ and assume that the amplitude U depends on the radial variable $r = \sqrt{x_1^2 + x_2^2}$. In this case, ordinary differential equations for the electromagnetic field amplitude and the plasma electron concentration are

$$\triangle_\perp U_{s,n} + [s + F(U_{s,n}^2)]U_{s,n} = 0\,, \tag{8.5}$$

$$F = -\frac{N_{s,n}}{\gamma_{s,n}}\,, \tag{8.6}$$

$$N_{s,n} = \max\{0, 1 + \triangle_\perp \gamma_{s,n}\}\,, \tag{8.7}$$

$$\gamma_{s,n} = \sqrt{1 + U_{s,n}^2}\,. \tag{8.8}$$

Here $\triangle_\perp = \partial_r^2 + r^{-1}\partial_r$ denotes the Laplacian with respect to r, and $N_{s,n}$ stands for the axially symmetrical plasma electron concentration function. The natural boundary conditions for problem (8.5)–(8.8) are

$$\frac{\mathrm{d}U_{s,n}}{\mathrm{d}r}(0) = 0\,, \qquad U_{s,n}(\infty) = 0\,. \tag{8.9}$$

The first condition assures axial symmetry of the solutions, and the second is necessary to provide for the convergence of the corresponding power integral. We designate these solutions as axially symmetrical relativistic and charge-displacement eigenmodes and explore their significance below.

The second of the boundary conditions given by (8.9) can be met when $0 < s < 1$. For such values of the problem parameter s, the dynamical system defined by (8.5)–(8.8) has three rest points; one of them is zero, and the corresponding localized axially symmetrical eigenmodes constitute a countable set that is ordered by the number of nodes n corresponding to finite values of the radial variable r [hence the index n in (8.5)–(8.8)] [7].

The zeroth (lowest) axially symmetrical eigenmode $U_{s,0}(r)$ [7,81], the first and second axially symmetrical relativistic and charge-displacement eigenmodes [7], and the corresponding electron densities given by (8.7), are depicted in Fig. 8.1 for $s = 0.95$. Since all these eigenmodes are calculated for the same value of the parameter s, one can conclude that cavitation may occur in the electromagnetic field–plasma systems described by higher eigenmodes even if it does not take place in such systems represented by lower eigenmodes.

Linear Stability of Axially Symmetrical Eigenmodes. As part of a general inquiry into relativistic and charge-displacement self-channeling stability, the linear stability of the lowest eigenmodes of problem (6.29)–(6.30) was treated in [97] with the help of the variational technique originally developed in [24–26] for quasi-linear Schroedinger equations. Assume that the modulus and the phase of the solution of (6.29)–(6.30) are given by

$$|a(r,z)| = U(r) + \alpha(r,z)\,,$$
$$\arg a(r,z) = sz + \beta(r,z)\,;$$

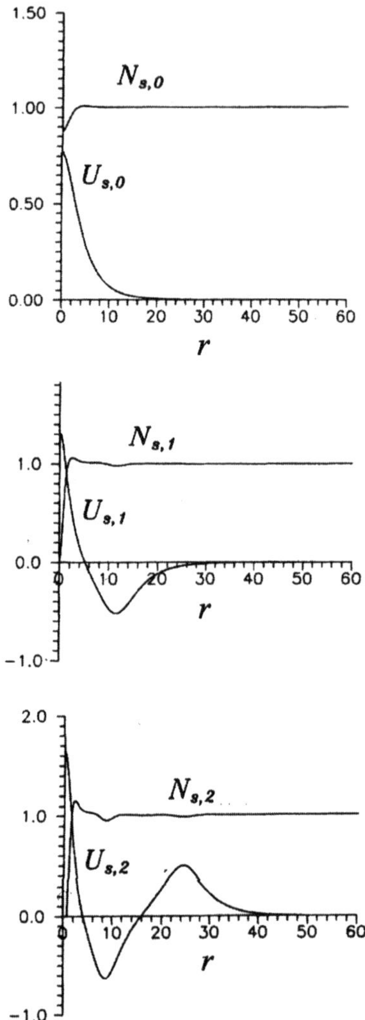

Fig. 8.1. The zeroth (lowest), first, and second axially symmetrical stationary relativistic and charge-displacement eigenmodes with $s = 0.95$: the normalized field amplitudes and the normalized electron concentration distributions [7]

α and β are small perturbations evolving against the eigenmode background. Substituting this solution in the dynamic problem (6.29)–(6.30) and linearizing the result, we find that

$$\alpha_t = L_0(r, \partial_r)\beta\,, \qquad \beta_t = L_1(r, \partial_r)\alpha\,,$$

$$L_0(r, \partial_r) = -\triangle_r + F[U_s^2(r)] - s + 1\,,$$

$$L_1(r, \partial_r) = -\Delta_r + s + 1 - \frac{U}{\gamma}\Delta_r \frac{U}{\gamma} F[U_s^2(r)] - \frac{1 + \Delta_r \gamma}{\gamma^3},$$

$$\gamma = \sqrt{1 + U^2}.$$

Obviously, $L_0 = L_0^* > 0$ and $L_1 = L_1^*$ in the space of square-integrable functions. Using the eigenfunction expansion

$$(\alpha, \beta)^T(r, z) = \sum_{n \in N+} (\alpha_n, \beta_n)^T(r) \exp(\mathrm{i}\kappa_n z),$$

one arrives at the eigenvalue problem,

$$L_0(r, \partial_r) L_1(r, \partial_r) \alpha_n = -\kappa_n^2 \alpha_n.$$

The stability of solutions of the form (8.1)–(8.4) can be inferred when $\kappa_n^2 > 0$ for all n. The evaluation of the eigenvalues of the operator $L_0(r, \partial_r)L_1(r, \partial_r)$ can be performed precisely following the procedure developed in [24–26]. According to these papers, the linear stability criterion is

$$\frac{\partial}{\partial s} \int_0^\infty U^2 r \, \mathrm{d}r < 0$$

For the lowest axisymmetric eigenmodes, the dependence of

$$\eta = \frac{P(s)}{P_{\mathrm{cr}}} = \int U_{s,0}^2 r \, \mathrm{d}r / P_{\mathrm{cr}}. \tag{8.10}$$

on s is presented in Fig. 8.2. Since, obviously, the above criterion is met, the lowest axisymmetric relativistic and charge-displacement eigenmodes of problem (6.29)–(6.30) are linearly stable to small perturbations [97].

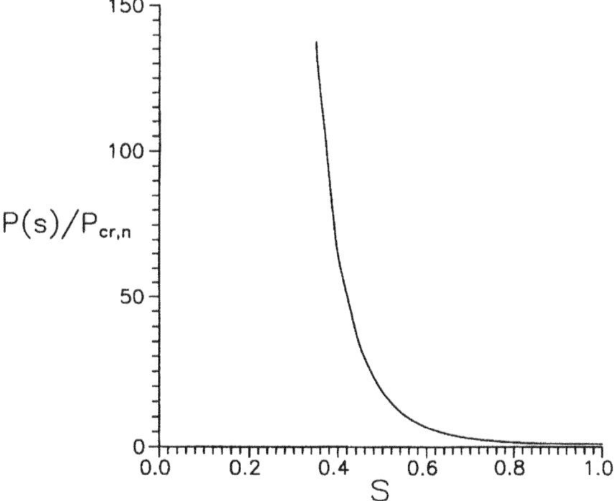

Fig. 8.2. Normalized power of the lowest axially symmetrical relativistic and charge-displacement eigenmode $U_{s,0}$ as a function of parameter s [97]

Note that the above procedure is performed assuming that the coefficients of the equations treated are continuous, which is the case when electron cavitation does not occur. A nonlinear study of relativistic and charge-displacement eigenmode stability without such additional assumptions is described in the following section of this chapter.

Relativistic and Charge-Displacement Self-Channeling Critical Power. It is a significant circumstance that the power

$$P_s = 2\pi \int U_{s,0}^2(r)\, r\, \mathrm{d}r$$

corresponding to the lowest eigenmode $U_{s,0}(r)$ intensity distribution depends on the parameter s (in contrast, the same quantity for the cubic Schroedinger equation's lowest axially symmetrical eigenmode is independent of s). This power shown in Fig. 8.2 is an evanescent function of s.

Obviously, the axially symmetrical eigenmodes discussed above are the simplest nonlinear solutions that demonstrate the possibility of relativistic and charge-displacement self-channeling, namely, of such regimes of superintense laser beam propagation into cold underdense plasmas when the optical field diffraction is suppressed by the relativistic and ponderomotive nonlinearity mechanisms resulting in spatially confined propagation modes. In this respect, a natural question is: What is the minimal power at which such regimes become possible? As shown in [7], this minimal power value obtained by using the lowest eigenmode partial solutions is of great significance for the dynamic theory, and in fact it represents adequately the relativistic and charge-displacement self-channeling critical power.

The calculation of the infimum of P_s by s in the interval $0 < s < 1$ is presented below, following [7]. This quantity, denoted by P_{cr},

$$P_{\mathrm{cr}} = \inf_s P_s$$

is called the critical power of relativistic and charge-displacement self-channeling. This power equals the critical power of purely relativistic self-focusing [7, 81]; the latter is a nonlinear phenomenon found in simulations using the purely relativistic nonlinearity model, which is obtained from that treated in this chapter by excluding the ponderomotive charge-displacement and freezing the electron concentration [94].

Since $U_{s,0}(r) \to 0$ and $n_{s,0}(r) \to 1$, for $s \to 1-0$, in this case [81],

$$\gamma_{s,0}^{-1} n_{s,0} - 1 \simeq (1 + U_{s,0}^2/2)^{-1} - 1 \simeq -U_{s,0}^2/2\,,$$

and

$$U_{s,0}(r) \simeq U_0(r)\,,$$

where $U_0(r)$ is the positive, monotonically decreasing (reaching zero for $r = \infty$ only) solution of the boundary-value problem,

$$\triangle_{\perp} U_0 - \delta U_0 + \frac{1}{2} g_0^3 = 0 \,,$$
$$\frac{dU_0}{dr}(0) = 0, \; U_0(\infty) = 0 \,,$$

with $\delta = 1 - s$. A change of variables which is typically used for treating cubic nonlinearity shows that

$$U_0(r) = \sqrt{2\delta}\, g_0(\sqrt{\delta} r) \,,$$

where g_0 is the wellknown Townes mode [18], i.e., the positive, monotonically vanishing solution of the following boundary-value problem:

$$\triangle_{\perp} g_0 - g_0 + g_0^3 = 0 \,,$$
$$\frac{dg_0}{dr}(0) = 0, \; g_0(\infty) = 0 \,.$$

It follows from the definition of P_{cr} and the above relation between U_0 and g_0 that

$$P_{\mathrm{cr}} = \inf_{0<s<1} P_s = 2\pi \int_0^\infty U_0^2(r)\, r\, \mathrm{d}r = 4\pi \int_0^\infty g_0^2(r)\, r\, \mathrm{d}r = 2P_{\mathrm{cr,c}} \,,$$

where

$$P_{\mathrm{cr,c}} = 2\pi \int_0^\infty g_0^2(r)\, r\, \mathrm{d}r$$

is the cubic medium self-focusing critical power [18, 31].

Finally, using the numerically calculated g_0, the critical power is [7]

$$P_{\mathrm{cr}} \simeq 23.4018 \,.$$

In dimensional units the relativistic and charge-displacement self-channeling critical power is [7]

$$P_{\mathrm{cr}} = \frac{m^2 c^5}{e^2} \int g_0^2(r) r\, \mathrm{d}r \left(\frac{\omega}{\omega_{\mathrm{p}}}\right)^2 \simeq 1.6198 \times 10^{10} \left(\frac{\omega}{\omega_{\mathrm{p}}}\right)^2 \mathrm{W} \,. \tag{8.11}$$

8.2 General Sufficient Condition for Relativistic and Charge-Displacement Self-Channeling

It is wellknown that the sufficient condition for a beam self-trapping in a medium with a saturable nonlinearity is that the Hamiltonian of the corresponding Schroedinger equation must be negative [31]. Chen and Sudan [82] established a similar criterion for relativistic and charge-displacement self-channeling using the nonlinear Schroedinger equation model (6.29)–(6.30).

Let us assume that no electron cavitation occurs and consider the integrals of motion given by (6.31) and (6.32). Obviously,

$$N \max |a|^2 \geq \int |a|^4\, \mathrm{d}^2 x_{\perp} \,,$$

and $|a|^4 \geq 4(\gamma-1)^2$, whereas the integral (6.31) can be rewritten in the form

$$H = N + \int \left[(\nabla_\perp a, \nabla_\perp a^*) - |\nabla_\perp \gamma|^2 - (\gamma-1)^2 \right] \mathrm{d}^2 x_\perp ,$$

where the sum of the first and the second terms is nonnegative. Combining the above facts, one finds that

$$\max |a|^2 \geq \frac{-4(H-N)}{N} .$$

This inequality means that when the constant $H - N$, which represents the system's Hamiltonian, is negative, the maximal value of the laser radiation intensity $|a|^2$ remains greater than zero at all times. In other words, the diffraction of light cannot result in total diffusion of the laser beam propagating into a plasma and relativistic and charge-displacement self-channeling takes place.

8.3 Propagation of Axially Symmetrical Laser Beams in Cold Underdense Plasmas

Consider the nonstationary dynamics of relativistically intense laser beam propagation into cold underdense plasmas. As we shall see below, for a broad range of experimentally attainable conditions, relativistically intense laser beams evolve in plasmas in a remarkably interesting regime called relativistic and charge-displacement self-channeling.

8.3.1 Problem Formulation in Terms of Propagation Distance

Following paper [7] where relativistic and charge-displacement self-channeling was investigated, we use a somewhat different formulation of the model embodied in (6.29)–(6.30). First, the longitudinal spatial coordinate will play the role of the evolutionary variable instead of time, which can be accomplished by introducing variables (x_3, η) instead of (τ, η) in the derivations of Chap. 6. Second, it should be taken into account that in general the unperturbed plasma electron concentration is not necessarily uniform. Since the plasma is created by nonlinear ionization at the laser pulse's temporally leading edge and the radial intensity distribution is not uniform, the same may be true of the ionization stage and, consequently, the unperturbed plasma electron concentration. As a result, the laser pulse's central temporal part propagates into a plasma column.

Finally, following [7], a different normalization is adopted in the rest of this chapter. For numerical simulations, it is convenient to introduce normalized coordinates and the optical field amplitude as follows:

$$r_1 = r/r_0 , \qquad z_1 = z/(2k_0 r_0^2) , \qquad u(r_1, z_1) = a_0^{-1} a(r, z) ,$$

where r and z are the dimensional radial and longitudinal coordinates, r_0 is the initial intensity profile's characteristic radius (for a selected value of η), and $a_0 = \max|a(r,0)|$ (we adhere to the notations of [7]). Subscript "1" is omitted in what follows to simplify notation. The set of normalized equations governing the laser beam amplitude has the form [7],

$$u_z + \mathrm{i}\Delta_\perp u + \mathrm{i}F(r,|u|^2)u = 0, \ z > 0\,, \tag{8.12}$$

$$u(r,0) = u_0(r)\,, \qquad \max_r |u_0| = 1\,, \tag{8.13}$$

$$u_r(0,z) = 0\,, \qquad u(\infty,z) = 0\,. \tag{8.14}$$

Here, z stands for the longitudinal coordinate and plays the role of the evolutionary variable. The nonlinear term F is

$$F(r,I) = a_1\left[1 - \frac{\max\{0, f_1(r) + a_1^{-1}\Delta_\perp\sqrt{1+a_2 I}\}}{\sqrt{1+a_2 I}}\right], \tag{8.15}$$

and the dimensionless problem parameters a_1, a_2 are defined as

$$a_1 = (r_0 k_\mathrm{p})^2\,, \qquad a_2 = I_0/I_\mathrm{r}\,, \qquad I_0 = \frac{m^2\omega^2 c^3 a_0^2}{4\pi e^2}\,, \tag{8.16}$$

where I_0 is the maximal intensity at the medium entrance plane ($z = 0$) and I_r is the relativistic intensity.

The ratio of the power of the beam P_0 and the critical power of relativistic and charge-displacement self-focusing P_cr, defined in the previous section, is an important parameter that is largely responsible for the behavior of solutions of (8.12)–(8.15) and relativistically intense laser beam propagation regimes [7]. This ratio can be expressed as

$$\eta = \frac{P_0}{P_\mathrm{cr}} = \frac{a_1 a_2}{B}\int_0^\infty |u_0(r)|^2\, r\,\mathrm{d}r\,;$$

the dimensionless constant B is given by

$$B = 2\int_0^\infty g_0(r)^2 r\,\mathrm{d}r \simeq 3.72451\,.$$

When the initial transverse-intensity distribution is Gaussian, namely, $|u_0(r)|^2 = \exp(-r^2)$,

$$P_0/P_\mathrm{cr} = \frac{a_1 a_2}{2B}\,.$$

The high precision of the calculation of the magnitude of the relativistic and charge-displacement critical power requires a special explanation. Simulations reported in [7] showed that relativistic and charge-displacement self-channeling is a phenomenon of threshold character in the sense that laser beams whose powers are just slightly greater than P_cr undergo channeling. In this respect, the use of a rougher definition, for example, where $B = 4$, would result in wrong predictions in this respect. Even a laser pulse for which $P_0/P_{cr} = 1.0347$ propagates in the self-channeling regime.

Following [7], let us examine the propagation of relativistically intense laser beams with Gaussian and hyper-Gaussian initial transverse and longitudinal intensity distributions of the form,

$$I|_{z=0} = I_0(r,t) = I_m \exp\left[-(t/\tau)^{N_1} - (r/r_0)^{N_2}\right] , \tag{8.17}$$

where $N_1 \geq 2$ and $N_2 \geq 2$ (the above expression is written in dimensional variables). It is assumed that $I_m \simeq I_r \gg I^*=10^{16}\,\mathrm{W/cm^2}$; I^* is the approximate rapid nonlinear ionization threshold value [75, 76]. For an initially flat phase front, the spatial amplitude distribution of the incident radiation corresponding to (8.17) has the form

$$u_0(r) = \exp(-r^{N_2}/2) . \tag{8.18}$$

The dimensionless parameter a_2 in (8.16), corresponding to the incident pulse intensity $I_0(t)$ on the beam propagation axis (at $r = 0$), is

$$a_2 = I_0(t)/I_r = (I_m/I_r)\exp[(-t/\tau)^{N_1}] . \tag{8.19}$$

The transverse profile of the plasma column, created by the intense laser pulse front, can be simulated by the hyper-Gaussian function [7],

$$f(r) = \exp[(-r/r_*)^{N_3}] , \qquad N_3 \geq 2 , \tag{8.20}$$

where r is the radial coordinate. The plasma column aperture r_* can be estimated by using the relation [7]

$$I_0(r_*, t_0) \equiv I_0(t_0)\exp[-(r_*/r_0)^{N_2}] = I^* .$$

For example, for a Gaussian transverse intensity distribution ($N_2 = 2$), the aperture of the plasma column for $I^* = 10^{16}\,\mathrm{W/cm^2}$, $I_r \simeq 0.45\times 10^{20}\,\mathrm{W/cm^2}$, and $I_0(t_0) = 0.1 I_r$, is $r_* \simeq 2.47\, r_0$, which justifies the homogeneous plasma approximation $f(r) \equiv 1$ [7]. In contrast, it follows from the above equation that for plateaulike incident transverse intensity distributions, the aperture of the simulated plasma column r_* tends to the beam aperture r_0. In the example given above for $N_2 = 8$, $r_* \simeq 1.25\, r_0$. Thus, beam defocusing, which is significant because of the proximity of the laser beam and the plasma column apertures, should be expected to contribute substantially to the character of propagation of a laser pulse with a plateaulike incident transverse intensity distribution [7].

8.3.2 Relativistic and Charge-Displacement Self-Channeling

This section is dedicated to the description of relativistic and charge-displacement self-channeling, which is a unique phenomenon of laser energy self-concentration in matter and superintense electromagnetic radiation propagation in spatially confined modes. Numerical simulations show that relativistically intense, ultrashort ($\tau_{\rm ion} \gg \tau \gg \tau_{\rm e}$), axisymmetric laser pulses undergo self-channeling in plasmas under a broad range of conditions. Besides, it is established that a large fraction of the total incident power of the beam can be trapped in stabilized self-channeling modes confined to the propagation axis.

This paradigm is illustrated below by several examples. Typical results of a simulation of relativistic and charge-displacement self-channeling are exhibited in Fig. 8.3. In propagation of an initially Gaussian transverse wave form incident on a homogeneous plasma, with the parameters $\lambda = 0.248\,\mu$m, $I_0 = \frac{2}{3} I_{\rm r} = 3\times10^{19}\,$W/cm^2, $r_0 = 3\,\mu$m, $n_{e,0} = 7.5\times10^{20}\,$cm^{-3} ($a_1 = 248.6192$, $a_2 = 2/3$, $P_0/P_{\rm cr} = 22.252$), the numerical simulation demonstrates that, as soon as the first focus on the axis of propagation is formed, ponderomotive force causes electronic cavitation. *Complete expulsion* of the plasma electron component from the paraxial domain occurs (see Fig. 8.3b). This process results in a quasi-stabilized cavitated channel in the electron distribution which extends along the entire propagation axis past the first focus location [7]. The first focus involves approximately 45% of the total incident laser beam power. A fraction of the remaining power is "dissipated" via diffraction on the periphery, while some of it is temporarily involved in the formation of a pulsing ring-shaped structure (Fig. 8.3a). Subsequent energy exchange between this ring-shaped feature and the paraxial zone takes place, and as a consequence of this interaction, a certain part of the pulsing ring energy is diffracted away, whereas the remaining power returns to the paraxial domain [7].

The following finer details of self-channeling onset are revealed by simulations. As described above, after the formation of the first focus, considerable power is transferred from the paraxial focal zone to the ring-shaped intensity distribution, which, at this stage of the laser pulse evolution, contains approximately 68% of its total initial power. This intense ring, which spreads away from the paraxial domain, produces a corresponding ring-shaped cavity in the plasma electron component (see Fig. 8.3b). The refraction resulting from this strong modification of electron concentration, coupled with the relativistic self-focusing mechanism, causes the wave to return energy to the beam core. Thus, charge-displacement produces a potent additional self-focusing effect, which leads to the formation of a confined paraxial mode of high intensity stabilized along the axis of propagation. This phenomenon of confined propagation is called relativistic and charge-displacement self-channeling.

Note that the electrostatic energy associated with charge displacement is relatively small. Specifically, for the conditions of Fig. 8.3b (at $z = 95.4\,\mu$m), it accounts for only 0.18% of the total laser radiation energy per unit length.

The essential finding of these calculations is that the combined effect of relativistic and charge-displacement nonlinearities produces a strong tendency for the generation of spatially highly confined propagation modes which are *stabilized* along the propagation axis. Furthermore, studies performed for a range of parameters make it possible to conclude that these modes are ex-

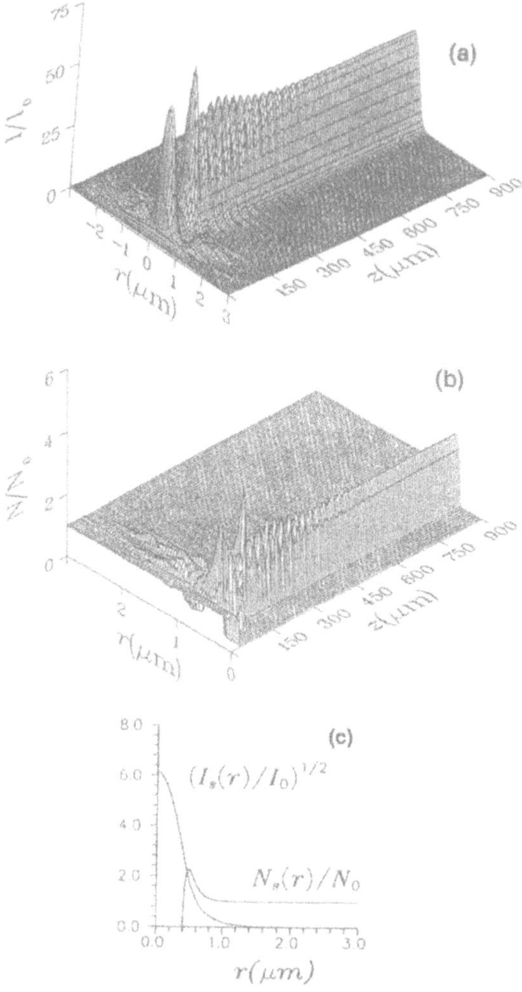

Fig. 8.3. The self-channeling of a laser pulse with a Gaussian initial transverse intensity distribution ($N_2 = 2$ in (8.18)) and a flat incident wave front in an initially homogeneous plasma. The beam and plasma parameter values are $I_0 = 3\times10^{19}$ W/cm^2, $r_0 = 3\,\mu$m, $\lambda = 248\,\mu$m, $n_{e,0} = 7.5\times10^{20}$ cm^{-3}. (**a**) The normalized intensity distribution; (**b**) the normalized electron concentration; (**c**) the radial dependence of the asymptotic normalized amplitude $\sqrt{I_s(r)/I_0}$ and electron concentration $N_s(r)/n_0$ distributions identified as the lowest relativistic and charge-displacement eigenmode with $s = 0.554$ [7]

ceptionally stable and that a considerable fraction of the incident power can be trapped in them. The basic result is the controlled generation of a very high maximal intensity in these channeled modes: the corresponding intensity magnitude for the range of parameters studied is 10^{21} W/cm^2 [7].

Important characteristics of the asymptotic behavior of these confined modes have also been established in [7], where it is shown that for large values of z, the amplitude distribution $u(r,z)$, tends asymptotically to the lowest eigenmode of the nonlinear Schroedinger equation. So, for the example discussed above, the asymptotic radial amplitude distribution corresponds to the lowest eigenmode with $s \simeq 0.554$. In this case, the asymptotic intensity distribution $I_s(r) = U_{s,0}^2(r)$ contains 46% of the total incident laser beam power. The normalized asymptotic field amplitude $\sqrt{I_s(r)/I_0}$ and the corresponding normalized asymptotic plasma electron density $N_s(r)/N_0$ are depicted in Fig. 8.3c. The charge displacement energy in the asymptotic state is approximately 10^{-3} of the total beam energy per unit length. Note that this tendency for a solution of a Schroedinger equation with a saturable nonlinearity to converge to the lowest stationary solution was originally discussed by Zakharov, Sobolev, and Synakh [31].

For the range of parameters studied, the calculations clearly show that charge displacement has a very strong influence on the character of the propagation after the emergence of the first focus. Although simulations with the frozen charge-displacement (purely relativistic nonlinearity) model yield pulsing waveguide propagation regimes for the same parameter values [7, 94], propagation governed by the combination of relativistic and charge-displacement nonlinearities results in stabilized and uniform channels.

Computations for the propagation of incident plateaulike wave forms with flat incident phase fronts in both homogeneous plasmas and plasma columns are also reported in [7] as well as studies of the behavior of focused Gaussian and plateaulike incident wave forms and defocused Gaussian incident wave forms in homogeneous plasmas. It is found that the main features described above, namely,

- relativistic and charge-displacement self-channeling of intense laser beams,
- propagation mode stabilization,
- confinement of a substantial fraction of incident laser beam power, and
- formation of paraxial cavitated channels in plasma electron fluid

are common aspects of superintense laser radiation dynamics in cold underdense plasmas over a wide range of physical conditions [7].

Let us consider several additional examples of relativistic and charge-displacement self-channeling. For the propagation of a beam with an incident plateaulike transverse intensity profile, given by (8.18) with $N_2 = 8$, and with an initially flat wave front incident on an initially homogeneous plasma (the values of the parameters are the same as above, and in this case $P_0/P_{\rm cr} = 20.168$), the intensity distribution asymptotic state contains approximately 77% of the incident power (see Fig. 8.4) [7]. The corresponding asymptotic amplitude is identified as the axially symmetrical relativistic and charge-displacement lowest eigenmode with $s \simeq 0.515$. The propagation of the same waveform in a plasma column, where the unperturbed electron concentration is defined by (8.20), where $N_3 = 8$ and $r_* = r_0$, results

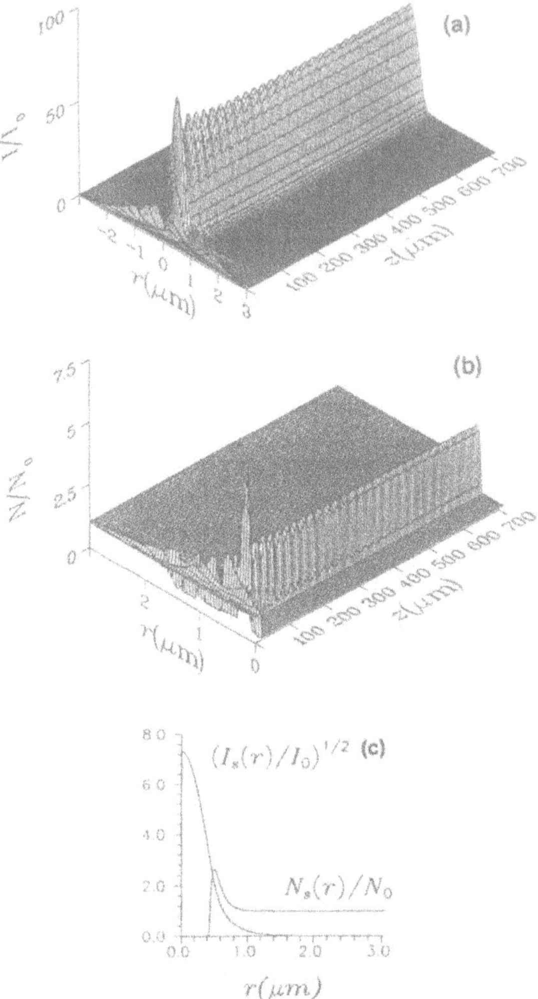

Fig. 8.4. The self-channeling of a laser pulse with a hyper-Gaussian initial transverse intensity distribution ($N_2 = 8$ in (8.18)) and a flat incident wave front in an initially homogeneous plasma. The beam and plasma parameter values are I_0=3×10^{19} W/cm^2, r_0=3 µm, λ=248 µm, $n_{e,0}$=7.5×10^{20} cm^{-3}. (**a**) The normalized intensity distribution; (**b**) The normalized electron concentration; (**c**) The radial dependence of the asymptotic normalized amplitude $\sqrt{I_s(r)/I_0}$ and electron concentration $N_s(r)/n_0$ distributions identified as the lowest relativistic and charge-displacement eigenmode with $s = 0.515$ [7]

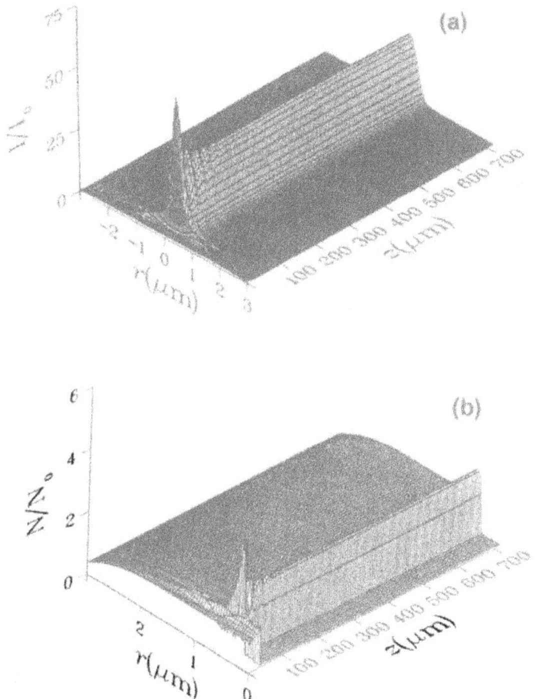

Fig. 8.5. The self-channeling of a laser pulse with a hyper-Gaussian initial transverse intensity distribution ($N_2 = 8$ in (8.18)) and a flat incident wave front in a preformed plasma column ($N_3 = 8$ and $r_* = r_0$ in (8.20)). The beam and plasma parameter values are I_0=3×10^{19} W/cm^2, r_0=3 μm, λ=248 μm, $n_{e,0}$=7.5×10^{20} cm^{-3}. (**a**) The normalized intensity distribution; (**b**) The normalized electron concentration [7]

in a quasi-stabilized intensity distribution containing 34% of the incident power (Fig. 8.5). Note that a model purely relativistic nonlinearity simulation involving the same plasma column also predicts the emergence of a quasi-stabilized intensity distribution but with only 25% of the incident beam power trapped; this provides evidence of the stabilizing effect of ponderomotive charge displacement. Overall, the computations reveal that ponderomotive charge-displacement, which typically results in electronic cavitation for sharp self-focusing, plays an important role in stabilizing the confined propagation mode that develops dynamically. This stabilization naturally occurs by refraction of the radiation into the central paraxial region. In fact, the influence of charge displacement on propagation is so strong that even *extremely* focused and defocused incident wave forms propagate in the self-channeling regime [7]. The latter fact means that self-channeling should be expected to be nearly independent of the initial laser beam focusing or defocusing.

The critical power P_{cr} for relativistic and charge-displacement self-channeling is defined in Sect. 8.1.2. Since, as we have seen above, relativistic and charge-displacement self-channeling is characterized by the dynamic emergence of the asymptotic transverse electromagnetic field amplitude and plasma electron concentration distributions corresponding to those given by the lowest eigenmodes of problem (8.12)–(8.15), and, according to the definition used, the critical power P_{cr} is the minimal power which can be "contained" in such eigenmodes, this quantity can be interpreted as the threshold power that separates asymptotic behavior, with respect to a large distance of propagation z, into two distinct classes (since this concept rests on an asymptotic property, it can be valid only in the limiting case of vanishing losses). For a power $P < P_{\text{cr}}$ the asymptotic transverse intensity distribution tends to zero at large z. In contrast, for $P > P_{\text{cr}}$, the asymptotic profile of the intensity tends to the lowest eigenmode of problem (8.12)–(8.15).

According to the results of [7], the relativistic and charge-displacement critical power for initially focused or defocused beams exceeds P_{cr} and depends on the initial wave front curvature. For a given wave front curvature magnitude, the critical power for the initially defocused beams is greater than that for those initially focused.

The principal result of this section is the finding that the combined action of relativistic and charge-displacement mechanisms can result in self-channeling with the formation of stabilized paraxial modes over a rather wide range of physical conditions. Moreover, these spatially confined modes involve corresponding cavitated channels in electron density. Finally, the characteristics of these channeled modes have asymptotic behavior that is described by the appropriate lowest eigenmodes of the governing nonlinear Schroedinger equation.

8.4 Filamentation Stability of Relativistic and Charge-Displacement Self-Channeling

As the next step, (8.12) and (8.15) are supplemented by the initial condition,

$$u(r,\varphi,0) = u_0(r,\varphi) \tag{8.21}$$

and the boundary conditions,

$$\lim_{r\to 0} r\frac{\partial u}{\partial r}(r,\varphi,z) = 0\,, \qquad u(\infty,\varphi,z) = 0\,,$$

$$u(r,\varphi+2\pi,z) = u(r,\varphi,z)\,.$$

To simulate initial azimuthally dependent perturbations, the following model expression is used in [97] to formulate the initial condition (8.21) for (8.12):

$$u_0(r,\varphi) = V_0(r)\left(1+\sum_{q=1}^{n}\varepsilon_q\cos q\varphi\right)\,, \tag{8.22}$$

where the unperturbed initial beam amplitude can be defined, for example, as

$$V_0(r) = \exp\left(-\frac{r^{N_2}}{2}\right) . \tag{8.23}$$

A range of parameters of the azimuthally dependent perturbations in (8.22) is examined. The following dimensionless quantity can serve as a measure of the initial azimuthal perturbation's magnitude:

$$\delta_{\text{int}} = \max_{r,\varphi_1,\varphi_2} |I_0(r,\varphi_1) - I_0(r,\varphi_2)| . \tag{8.24}$$

As we shall see below, an important role in characterizing the behavior of relativistically intense laser beams with violated axial symmetry is played by the following two parameters derived from the primary physical parameters λ, r_0, P_0, and $n_{e,0}$:

$$\eta = \frac{P_0}{P_{\text{cr}}} = \frac{a_1 a_2}{P_{cr,n}} \int_0^\infty \int_0^{2\pi} |u_0|^2 r \, \mathrm{d}r \, \mathrm{d}\varphi , \tag{8.25}$$

and

$$\begin{aligned} \rho_0 &= \sqrt{\frac{1}{\pi I_0} \int_0^\infty \int_0^{2\pi} I_0(r,\varphi) r \, \mathrm{d}r \, \mathrm{d}\varphi} \, \frac{\omega_{p,0}}{c} \\ &= \sqrt{\frac{1}{\pi} \int_0^\infty \int_0^{2\pi} |u_0(r,\varphi)|^2 r \, \mathrm{d}r \, \mathrm{d}\varphi} \, \frac{r_0 \omega_{p,0}}{c} , \end{aligned} \tag{8.26}$$

related to the laser beam power and initial radius, respectively. In axial symmetry,

$$\rho_0 = \frac{r_0 \omega_{\text{p},0}}{c} = \sqrt{a_1} . \tag{8.27}$$

A concept of relativistic and charge-displacement self-channeling stability to filamentation is introduced in [97] specifically for the range of phenomena discussed in this chapter. Since the essential aspect of relativistic and charge-displacement self-channeling is the unique nonlinear laser beam energy self-concentration associated with this effect, the stability of the beam power confinement must be reflected in stability definition rather than the magnitude of the azimuthal displacement of the laser beam filaments caused by axial symmetry violation. Following the above paper, let us consider relativistically intense laser beam propagation in a cold underdense plasma stable when it results in the emergence of a single main channel that (1) contains power P_{c} substantially exceeding the relativistic and charge-displacement self-channeling critical power P_{cr} (typically 30–50% of the incident laser beam power P_0) and (2) $\mathrm{d}P_{\text{c}}/\mathrm{d}P_0 > 0$. The unstable regime which can also be termed strong filamentation is characterized by the formation of multiple channels whose powers are comparable to P_{cr}.

A study of the stability to filamentation in initially homogeneous ($f(r) \equiv 1$) plasmas and preformed plasma columns is described below, following [97].

Basically, simulations show that if $P_0 > P_{cr}$, channeled propagation develops which is in many respects similar to that in the axisymmetric case. However, the detailed propagation pattern can differ significantly from the axisymmetric situation. Overall, the following features of intense laser beam evolution in a cold underdense plasmas in the presence of azimuthally dependent perturbations are observed in simulations [97]:

1. A propagation regime exists, in which the formation of a single central or slightly radially displaced channel occurs that contains a substantial fraction (over 30–50%) of the incident laser beam power.
2. A regime exists with one main channel containing a substantial fraction of the incident power (nominally over 30–50%) and one or more subsidiary channels, each of which contains power $\approx P_{cr}$.
3. A strongly unstable regime exists, characterized by the formation of several filamentary channels each of them containing power $\approx P_{cr}$.

In terms of the stability definition given above, regimes (1) and (2) are considered stable, whereas regime (3) is unstable. The definition of stability requires that an increase in the incident power leads to an increase in the power confined in the channel with the highest power. The transition from stable self-channeling to the unstable regime is characterized by the opposite tendency, specifically, an increase in the incident power causes a corresponding decrease in the power confined. In the unstable propagation regime (3), the number of channels tends to increase as the incident power increases, but no increase occurs in the power trapped in an individual channel. Therefore, a further increase in the incident power in the unstable domain results only in the emergence of a larger number of peripheral filamentary channels, and the power diffuses into an expanding volume of space.

8.4.1 Eigenmode Stability to Filamentation

Since the concept of relativistic and charge-displacement self-channeling in mathematical terms rests on the property of the field–plasma systems to evolve dynamically toward the lowest eigenmodes of problem (6.29)–(6.30), the stability of these eigenmodes to filamentation is an issue of high importance for the overall understanding of the relativistic and charge-displacement effect. According to the conclusions of [97], the channeled propagation associated with lowest eigenmodes is extremely stable to azimuthal perturbations, even for very high incident powers. In this paper, initial laser beam amplitude distributions of the form (8.22) were considered where the function V_0 is given by

$$V_0(r) = U_{s,0}(r)/U_{s,0}(0)\,, \qquad r = r'/\rho_{e,0}\,;$$

$U_{s,0}$ is the lowest relativistic and charge-displacement axially symmetrical eigenmode (specifically, the lowest eigenmode with $s = 0.4$ is considered), and $\rho_{e,0}$ is its radius defined as

$$\rho_{e,0} = \sqrt{2\int_0^\infty U_{s,0}^2(\rho)\rho d\rho / U_{s,0}(0)}\,.$$

For $s = 0.4$, $\rho_{e,0} = 2.0067$ and $\eta = 65.071$. In one of the cases treated in [97], the values of the parameters characterizing the perturbation magnitude were $\varepsilon_q = 10^{-2}$, $q = \overline{1,4}$, $\delta_{int} = 6.05 \times 10^{-2}$. A stable channel develops rapidly where essentially all of the incident laser beam power is trapped. Moreover, additional calculations have shown [97] that this stable response is maintained for both higher powers ($\eta < 124$) and for higher initial azimuthal perturbation levels ($\delta_{int} \leq 1.38$). The stability of the lowest eigenmodes is exceptionally solid.

The stability of the higher eigenmodes $U_{s,n}(r)$, $n \geq 1$, of problem (6.29)–(6.30) to filamentation is considered in [97] as well. Computations have shown that self-channeling in these modes can also be stable despite the presence of small azimuthally dependent perturbations in the initial conditions. In particular, stability has been inferred for initial amplitude distributions corresponding to the first eigenmode $U_{s,1}(r)$ with $s = 0.4$ when the perturbation magnitude is $\delta_{int} = 0.1953$. Note that for this eigenmode, the power parameter η and the dimensionless radius are 412.07 and 2.7743, respectively.

In contrast to the lowest eigenmode case, the stable self-channeling of pulses with transverse intensity distributions corresponding to azimuthally perturbed higher eigenmodes is not an obviously expected outcome. The ponderomotive radial charge displacement of electrons that leads to the formation of cavitated zones in the plasma electron fluid is the main factor responsible for the observed stability.

8.4.2 Stability of Initially Hyper-Gaussian and Gaussian Beams in Initially Homogeneous Plasmas

The boundary separating the stable and unstable regimes in the (η, ρ_0) plane is developed by simulating the powerful laser–plasma interactions for a wide range of initial conditions. The results for an initially homogeneous plasma and hyper-Gaussian beams [with $N_2 = 8$ in (8.23)] and with $\delta_{int} = 0.048$ are illustrated in the filamentation stability map shown in Fig. 8.6. Obviously, a large zone of stable propagation exists.

Figures 8.7 a and b, respectively, display the normalized laser radiation intensity and the corresponding plasma electron concentration for a perturbed incident waveform with $\varepsilon_q = 10^{-2}$, $q = \overline{1,4}$ in (8.22). In particular, this figure represents the case where $\lambda = 248\,\text{nm}$, $P_0 = 2 \times 10^{12}\,\text{W}$, $r_0 = 3.5\,\mu\text{m}$, and $n_{e,0} = 1.35 \times 10^{21}\,\text{cm}^{-3}$, values that can be experimentally attained and pertain to point A1 in Fig. 8.6. A single channel slightly shifted from the z axis forms rapidly. The power trapped in it is $\approx 6.1\,P_{cr}$, which is approximately two-thirds of that incident. An increase in the incident laser beam power P_0 leads to a corresponding increase in the level of power P_c trapped in the channel, so that $dP_c/dP_0 > 0$. This fact is exemplified in Fig. 8.8,

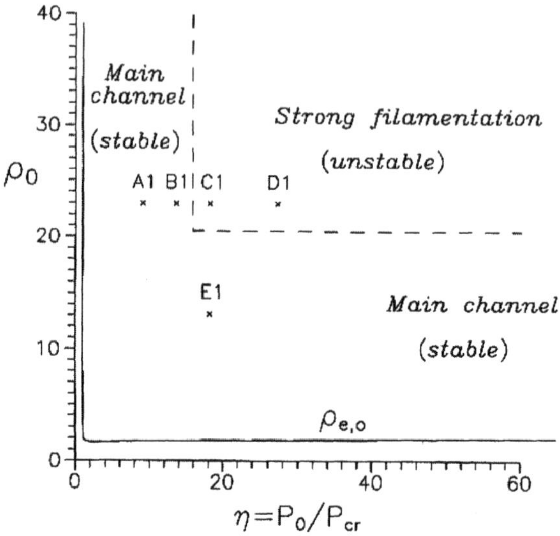

Fig. 8.6. Filamentation stability map for relativistic and charge-displacement self-channeling of initially hyper-Gaussian intense laser beams ($N_2 = 8$) in initially homogeneous plasmas. The azimuthally dependent perturbation's magnitude is $\delta_{\mathrm{int}} = 0.048$, ($\epsilon = 0.01$, $q = 1 - 4$). Stable and unstable regions of initial conditions in the (η, ρ_0) and the relation $\rho_{\mathrm{e},0}(\eta)$ for the lowest axially symmetrical relativistic and charge-displacement eigenmodes are shown. The points A1–E1 correspond to specific simulations presented in the following figures [97]

where stable relativistic and charge-displacement self-channeling is shown for a perturbed laser beam ($\lambda = 0.248\,\mu\mathrm{m}$, $P_0 = 4 \times 10^{12}\,\mathrm{W}$, $r_0 = 3.5\,\mu\mathrm{m}$, and $n_{\mathrm{e},0} = 1.35 \times 10^{21}\,\mathrm{cm}^{-3}$) with $\eta = 13.67$ and $\rho_0 = 23.007$. This case, denoted by point B1 in Fig. 8.6, corresponds to $\lambda = 0.248\,\mu\mathrm{m}$, $P_0 = 3 \times 10^{12}\,\mathrm{W}$, $r_0 = 3.5\,\mu\mathrm{m}$, and $n_{\mathrm{e},0} = 1.35 \times 10^{21}\,\mathrm{cm}^{-3}$. For these conditions, the resulting channel contains power $\approx 9P_{\mathrm{cr}}$.

A sufficiently high level of initial power can lead to propagation instability and strong filamentation. Such an example is shown in Fig. 8.9a–f, where the normalized transverse laser radiation intensity distributions for several different values of z are depicted. In this case, marked by point C1 in Fig. 8.6, $\eta = 18.23$ and $\rho_0 = 23.007$, values that correspond to $\lambda = 0.248\,\mu\mathrm{m}$, $P_0 = 4 \times 10^{12}\,\mathrm{W}$, $r_0 = 3.5\,\mu\mathrm{m}$, and $n_{\mathrm{e},0} = 1.35 \times 10^{21}\,\mathrm{cm}^{-3}$. Here, the beam disintegrates into several filaments. Although three filaments are stabilized in the central domain during the propagation, the most powerful of them carries only $5.2P_{\mathrm{cr}}$. It follows from the comparison with the previous example that an increase in the initial power P_0 has resulted in a decrease in the power trapped in the most powerful channel.

A further increase in P_0 for the same value of ρ_0 results in an increase in the number of filamentary channels. For example, Fig. 8.10 shows the asymptotic transverse intensity profile for $\eta = 27.34$ and $\rho_0 = 23.007$, values which

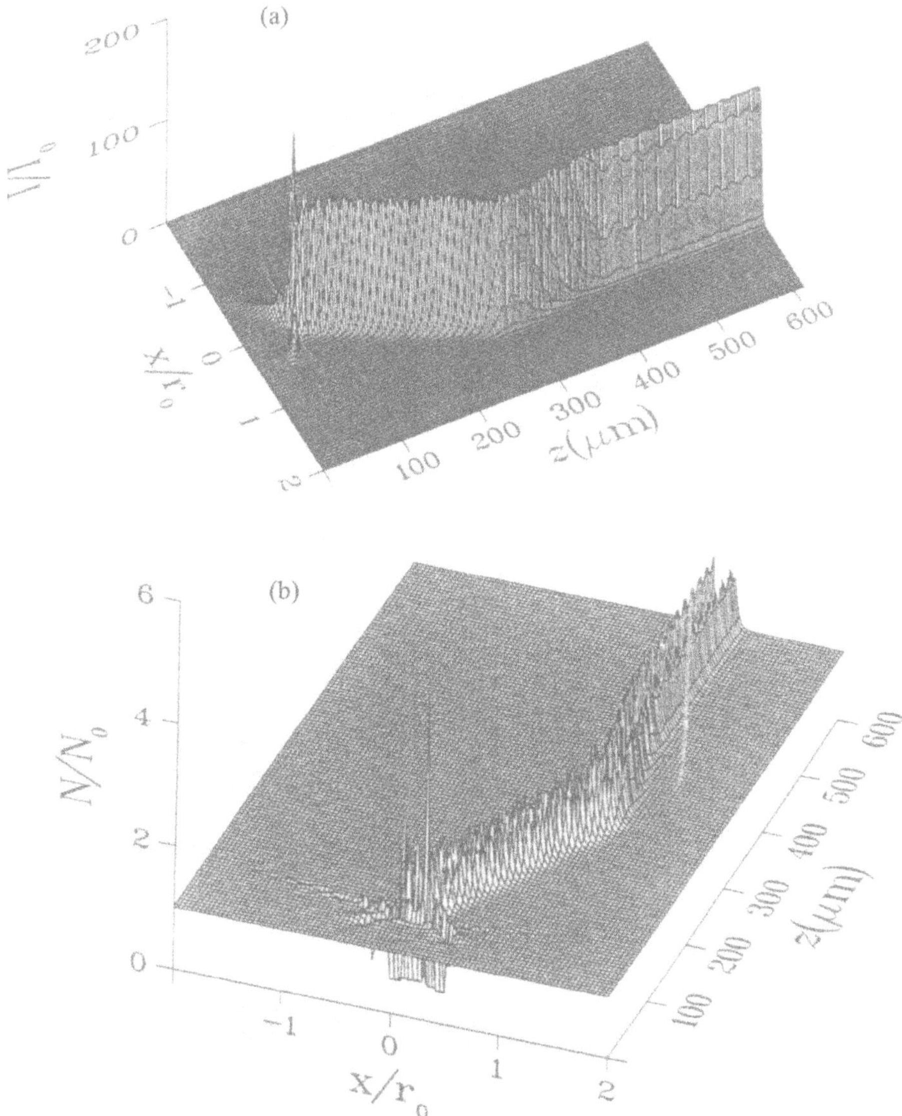

Fig. 8.7. Stable relativistic and charge displacement self-channeling of an azimuthally perturbed ($\epsilon = 0.01$, $q = 1-4$, and $\delta_{\rm int} = 0.048$) initially hyper-Gaussian intense laser beam ($N_2 = 8$) in an initially homogeneous plasma for the initial conditions corresponding to point A1 in the stability map shown in Fig. 8.6: $\eta = 9.113$ and $\rho_0 = 23.007$. The laser beam and plasma parameters are $\lambda = 248\,\rm nm$, $n_{e,0} = 1.35 \times 10^{21}\,\rm cm^{-3}$, $P_0 = 2 \times 10^{12}\,\rm W$, $r_0 = 3.5\,\mu m$, and $I_0 = 5.7 \times 10^{18}\,\rm W/cm^2$. Coordinate z represents the propagation direction, and x is one of the transverse variables. (**a**) Normalized intensity distribution; (**b**) Normalized electron concentration [97]

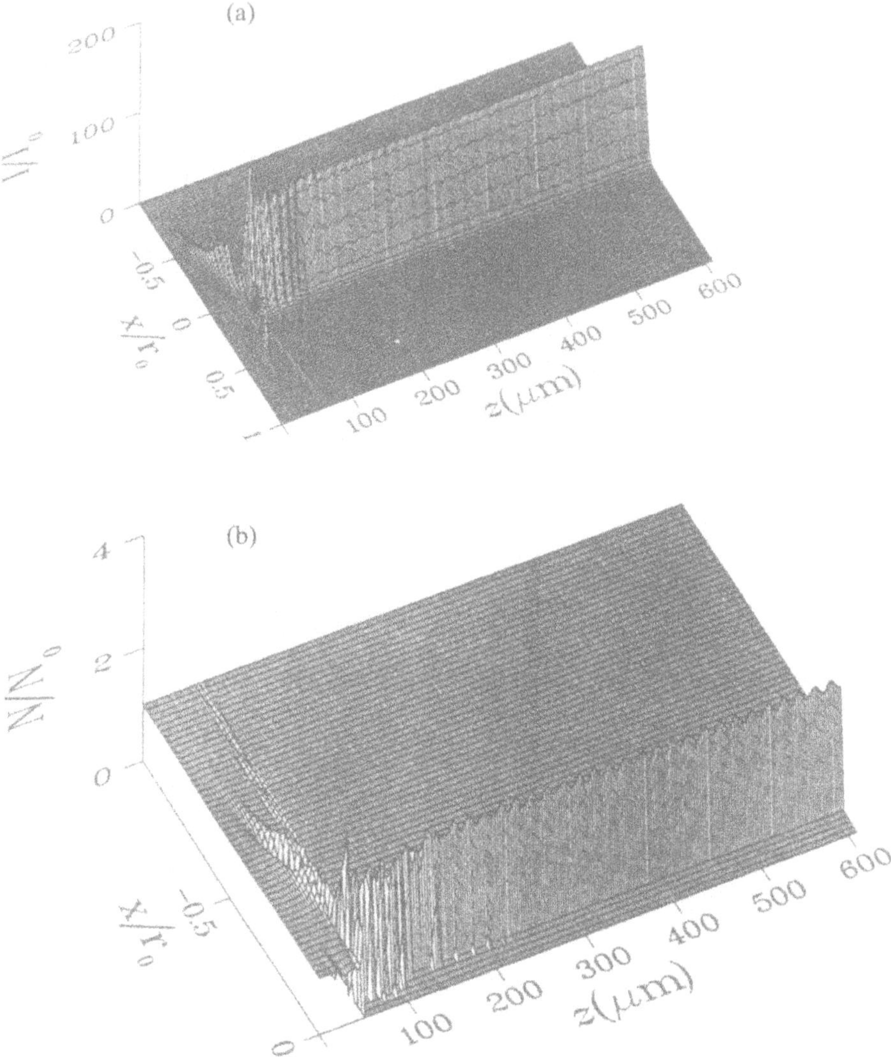

Fig. 8.8. Stable relativistic and charge displacement self-channeling of an azimuthally perturbed ($\epsilon = 0.01$, $q = 1 - 4$, and $\delta_{\text{int}} = 0.048$) initially hyper-Gaussian intense laser beam ($N_2 = 8$) in an initially homogeneous plasma for the initial conditions corresponding to point B1 in the stability map shown in Fig. 8.6: $\eta = 13.67$ and $\rho_0 = 23.007$. The laser beam and plasma parameters are $\lambda = 248\,\text{nm}$, $n_{\text{e},0} = 1.35 \times 10^{21}\,\text{cm}^{-3}$, $P_0 = 3 \times 10^{12}\,\text{W}$, $r_0 = 3.5\,\mu\text{m}$, and $I_0 = 8.6 \times 10^{18}\,\text{W/cm}^2$. Coordinate z represents the propagation direction, and x is one of the transverse variables. (**a**) Normalized intensity distribution; (**b**) Normalized electron concentration [97]

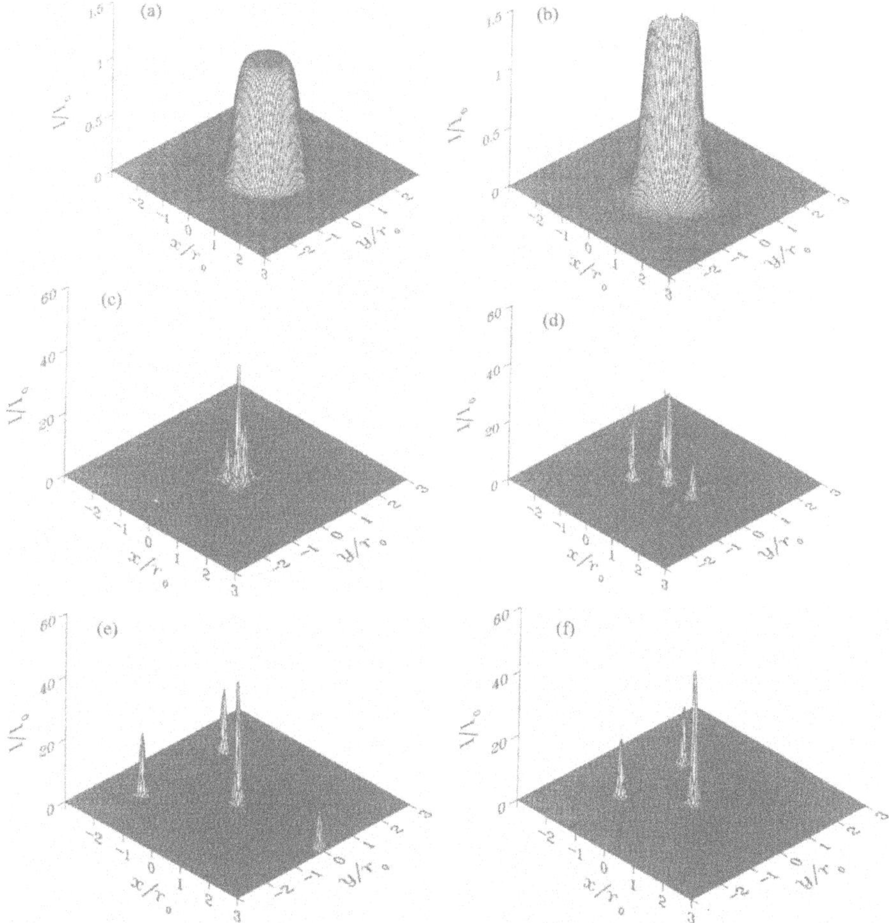

Fig. 8.9. Unstable relativistic and charge displacement self-channeling of an azimuthally perturbed ($\epsilon = 0.01$, $q = 1 - 4$, and $\delta_{\rm int} = 0.048$) initially hyper-Gaussian intense laser beam ($N_2 = 8$) in an initially homogeneous plasma for the initial conditions corresponding to point C1 in the stability map shown in Fig. 8.6: $\eta = 18.22$ and $\rho_0 = 23.007$. The laser beam and plasma parameters are $\lambda = 248\,\text{nm}$, $n_{\rm e,0} = 1.35 \times 10^{21}\,\text{cm}^{-3}$, $P_0 = 4 \times 10^{12}\,\text{W}$, $r_0 = 3.5\,\mu\text{m}$, and $I_0 = 1.2 \times 10^{19}\,\text{W/cm}^2$. Laser beam filamentary structure transverse cross sections at z=0, 12.25, 49.0, 85.75, and 171.5 µm, and the stabilized transverse asymptotic laser radiation intensity distribution are shown [97]

pertain to point D1 on the stability map shown in Fig. 8.6. These values represent $\lambda = 0.248\,\mu\text{m}$, $P_0 = 4\times 10^{12}\,\text{W}$, $r_0 = 3.5\,\mu\text{m}$, and $n_{\rm e,0} = 1.35\times 10^{21}\,\text{cm}^{-3}$. Comparison of the asymptotic transverse profiles presented in Figs. 8.9f and 8.10 shows that increasing the initial power, for the same ρ_0, simply results in the formation of an additional filament.

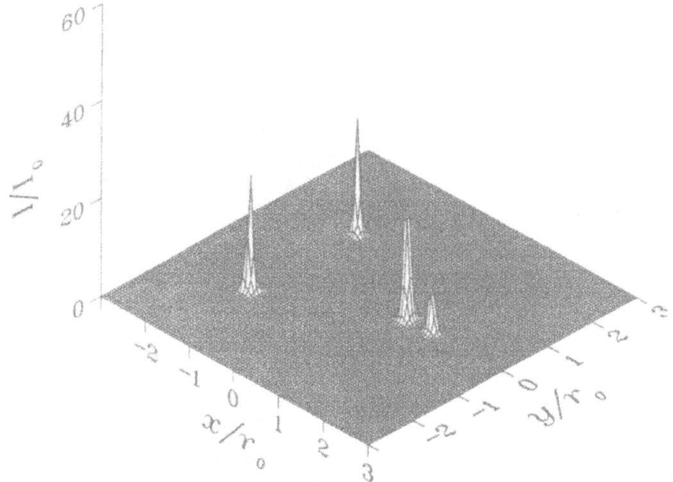

Fig. 8.10. Unstable relativistic and charge displacement self-channeling of an azimuthally perturbed ($\epsilon = 0.01$, $q = 1 - 4$, and $\delta_{\rm int} = 0.048$) initially hyper-Gaussian intense laser beam ($N_2 = 8$) in an initially homogeneous plasma for the initial conditions corresponding to point D1 in the stability map shown in Fig. 8.6: $\eta = 9.113$ and $\rho_0 = 27.34$. The laser beam and plasma parameters are $\lambda = 248\,\mathrm{nm}$, $n_{\rm e,0} = 1.35 \times 10^{21}\,\mathrm{cm}^{-3}$, $P_0 = 6 \times 10^{12}\,\mathrm{W}$, $r_0 = 3.5\,\mu\mathrm{m}$, and $I_0 = 1.7 \times 10^{19}\,\mathrm{W/cm}^2$. The filamentary structure of the stabilized asymptotic transverse normalized laser radiation intensity distribution is shown [97]

The high relativistic and charge-displacement self-channeling stability associated with eigenmodes indicates that the following concept for providing for the stability of this type of regime can be effective. In terms of providing favorable conditions for the laser energy self-concentration in matter via the self-channeling regime, a waveform as close as possible to the lowest relativistic and charge-displacement eigenmode should be the optimal input for the system comprising a relativistically intense coherent electromagnetic field and a cold underdense plasma.

In [97], the pair of quantities (η, ρ_0) (the power and the amplitude distribution radius) was proposed as the fitting parameters to characterize the proximity of an incident laser radiation waveform and an eigenmode. In other words, stable relativistic and charge-displacement self-channeling can be reached by arranging the initial conditions, so that $\eta = P_0/P_{\rm cr}$ and ρ_0 are close to values corresponding to the lowest eigenmodes $U_{s,0}$ of the system. Therefore, the tendency to favor conditions closer to the eigenmode can be illustrated by comparing the behavior corresponding to point C1 in Fig. 8.6 with that corresponding to point E1 located below it. The latter point is characterized by $\eta = 18.23$ and $\rho_0 = 13.111$, parameters that correspond to $\lambda = 0.248\,\mu\mathrm{m}$, $P_0 = 4 \times 10^{12}\,\mathrm{W}$, $r_0 = 2\,\mu\mathrm{m}$, and $n_{\rm e,0} = 1.35 \times 10^{21}\,\mathrm{cm}^{-3}$. The evolution of the pertinent waveforms is illustrated in Fig. 8.11. The propaga-

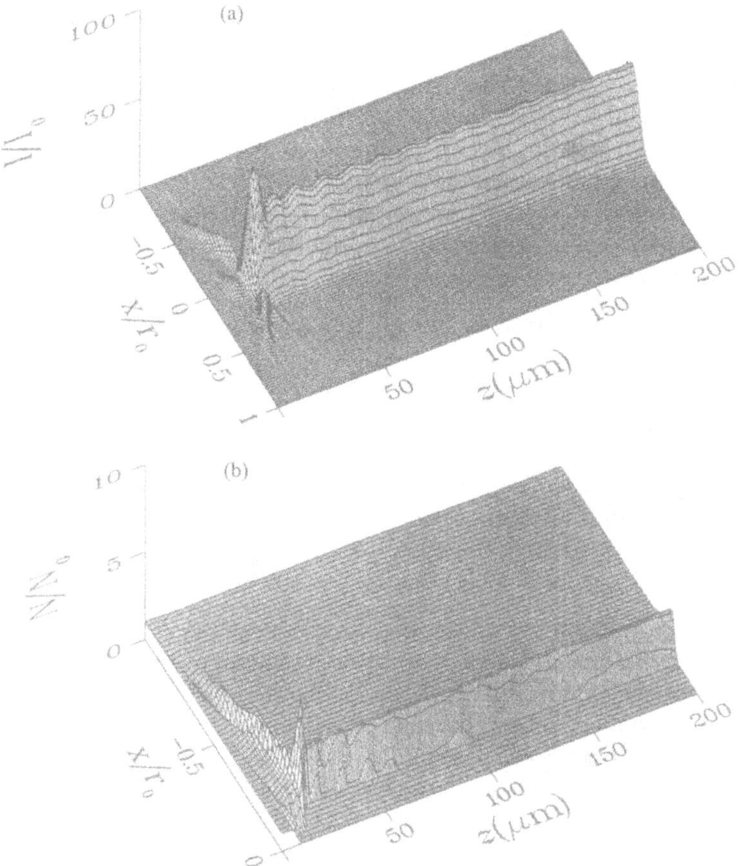

Fig. 8.11. Stable relativistic and charge-displacement self-channeling of an azimuthally perturbed ($\epsilon = 0.01$, $q = 1 - 4$, and $\delta_{\text{int}} = 0.048$) initially hyper-Gaussian intense laser beam ($N_2 = 8$) in an initially homogeneous plasma for the initial conditions corresponding to point E1 in the stability map shown in Fig. 8.6: $\eta = 18.23$ and $\rho_0 = 13.147$. The laser beam and plasma parameters are $\lambda = 248\,\text{nm}$, $n_{e,0} = 1.35 \times 10^{21}\,\text{cm}^{-3}$, $P_0 = 4 \times 10^{12}\,\text{W}$, $r_0 = 2\,\mu\text{m}$, and $I_0 = 3.5 \times 10^{19}\,\text{W/cm}^2$. Coordinate z represents the propagation direction, and x is one of the transverse variables. (**a**) Normalized intensity distribution; (**b**) Normalized electron concentration [97]

tion results in the formation of a single axial channel in which the confined power is $\approx 14P_{\text{cr}}$, the majority of that incident. Therefore, a decrease in the value of ρ_0 in moving from point C1 to point E1 located closer to the eigenmode in the (η, ρ_0) space restores stable behavior.

Simulations demonstrate that very high laser beam intensity can be associated with stable channeled propagation for initial conditions that fall sufficiently close to the eigenmode. For example, in one of the simulations described in [97], stable relativistic and charge-displacement self-channeling is observed for $\eta = 145.81$, $\rho_0 = 6.5734$, $\varepsilon_q = 10^{-2}$, $q = \overline{1,4}$, values corresponding to $\lambda = 0.248\,\mu\text{m}$, $P_0 = 3.2 \times 10^{13}\,\text{W}$, $r_0 = 1\,\mu\text{m}$, $n_{e,0} = 1.35 \times 10^{21}\,\text{cm}^{-3}$, and $\delta_{\text{int}} = 5 \times 10^{-2}$. The single channel arising in this case contains power $\approx 128P_{\text{cr}}$, a large fraction of that incident, and the laser beam intensity on the channel axis stabilizes at a level of approximately 10^{22}W/cm^2, a value corresponding to a maximal electric field greater than $500\,e/a_0^2$.

The study of the stability of the propagation of initially hyper-Gaussian relativistically intense laser beams in initially homogeneous cold underdense plasmas makes it possible to conclude the following [97]:

1. Large zones of stability exist in the (η, ρ_0) plane for $\eta > 1$ and $\rho_0 > 0.5\rho_{e,0}$.
2. Relativistic and charge-displacement self-channeling remains stable to filamentation, even at high levels of azimuthally dependent perturbations.
3. The lowest eigenmodes $U_{s,0}$ describe particularly stable types of relativistically intense electromagnetic waves.
4. The optimal efficiency of laser beam power confinement based on the relativistic and charge-displacement self-channeling effect is provided for by launching the wave matched to the lowest eigenmode.

A comparison of the behavior of initially Gaussian pulses with that of the initially hyper-Gaussian pulses makes it possible to examine the impact of radial gradients on overall relativistic and charge-displacement self-channeling dynamics and, in particular, on self-channeling stability to filamentation. Let the initial amplitude distribution be given by (8.22) and (8.23) with $N_2 = 2$, $\varepsilon_q = 0.01$, $q = \overline{1,4}$. In this case, the perturbation level makes $\delta_{\text{int}} = 6.3\times10^{-2}$. The corresponding filamentation stability map is shown in Fig. 8.12. Comparison of the maps shown in Figs. 8.6 and 8.12 demonstrates that the overall structure of the stable region is not strongly affected by modification of the incident transverse profiles. The stable zone in the map for Gaussian pulses is slightly smaller, but the contours of this region largely remain the same.

A comparison of two cases that are close in terms of the parameters η and ρ_0 leads to the conclusion that the waveforms evolving out of both initially hyper-Gaussian and initially Gaussian transverse intensity distributions exhibit significant deviations from axial symmetry, develop channels with a confined power considerably greater than P_{cr}, and satisfy all of the filamentation stability criteria formulated above [97].

Numerous specific examples presented in [97] indicate that the tendency to develop stable confined relativistic and charge-displacement self-channeling modes is not sensitive to the incident radiation radial profile.

8.4.3 Filamentation Stability in Preformed Plasma Columns

As noted above, when a plasma is dynamically formed by the front of a propagating superintense laser pulse, the electron concentration transverse distribution can be substantially inhomogeneous. For this reason, the issue of

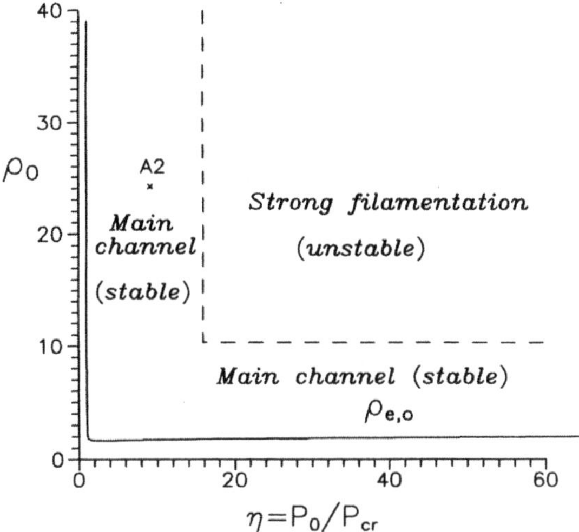

Fig. 8.12. Filamentation stability map for the relativistic and charge-displacement self-channeling of initially Gaussian intense laser beams ($N_2 = 2$) in initially homogeneous plasmas. The azimuthally dependent perturbation's magnitude makes $\delta_{\text{int}} = 0.063$, ($\epsilon = 0.01$, $q = 1-4$). Stable and unstable regions of initial conditions in the (η, ρ_0) and the relation $\rho_{e,0}(\eta)$ for the lowest axially symmetrical relativistic and charge-displacement eigenmodes are shown [97]

the stability of relativistically intense laser beam propagation in preformed plasma columns is also addressed in [97]. Overall, the conclusion of this paper is that for a sufficiently high initial laser beam power, confined propagation stable to filamentation is possible in preformed plasmas columns, as well as in initially homogeneous plasmas.

In the following example, the plasma column's initial transverse profile is modeled by a hyper-Gaussian function defined by (8.20) with $N_3 = 8$, assuming that the initial radii of the beam and the plasma column are equal, $r_* = r_0$, and that the incident wave has a hyper-Gaussian transverse intensity distribution and a flat phase font. The initial azimuthally perturbed amplitude distribution is given by (8.22) and (8.23) with $N_2 = 8$, $\varepsilon_q = 0.03$, $q = \overline{1,4}$. The corresponding perturbation level makes $\delta_{\text{int}} = 0.1953$.

Since the initial radii of the beam and plasma column are sufficiently close, substantial defocusing occurs at the initial stage of the laser–plasma interaction, an effect which results in "dissipation" of a fraction of the laser beam power at the periphery. Therefore, the power trapped in any resulting channel is smaller than it would have been if the plasma had been initially homogeneous. However, due to the rapid readjustment of the propagating beam's transverse profile that can occur in the column, propagation stabilization occurs more quickly in this case. The threshold power of self-channeling in plasma columns P_{col} is greater than the power P_{cr} for self-channeling in an initially homogeneous plasmas. The power P_{col} is the minimal initial power necessary for an extended uniform laser channel to emerge.

The value of P_{col} is determined by the initial transverse profile and radius of the plasma column. For $N_3 = 8$ and $r_* = r_0$, the estimated value of relativistic and charge-displacement self-channeling in a preformed plasma column is $P_{col} = 5.1 P_{cr}$. Naturally, the magnitude of P_{col} is lower for higher values of r_* since the situation tends physically to the homogeneous case in the $r_* \gg r_0$ limit [94].

An important circumstance observed in [97] is that, when $r_* \approx r_0$, the strong filamentation regime is significantly altered by the defocusing of peripheral filaments. Peripheral channels are defocused at the plasma column edge and consequently, the power within them is distributed over a wide region.

The results established in [97] are embodied in the map of filamentation stability shown in Fig. 8.13. As one can easily see, the general features of the stable domain are qualitatively the same as in the map for initially homogeneous plasmas depicted in Figs. 8.6. Readers can find further details in [97].

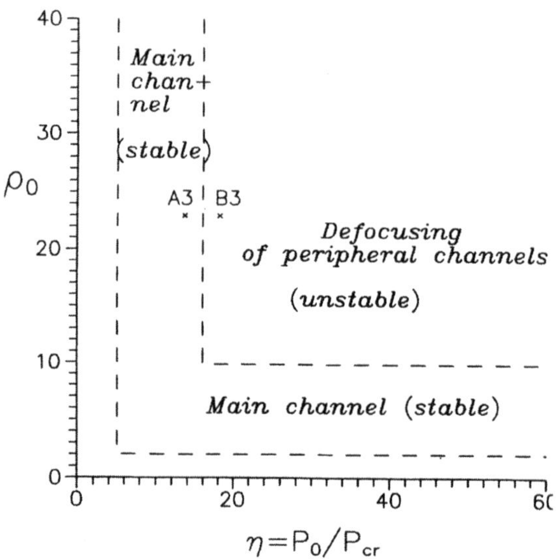

Fig. 8.13. Filamentation stability map for the relativistic and charge-displacement self-channeling of initially hyper-Gaussian intense laser beams ($N_2 = 8$) in preformed plasma columns ($N_3 = 8$ and $r_* = r_0$). The azimuthally dependent perturbation's magnitude makes $\delta_{int} = 0.1458$, ($\epsilon = 0.03$, $q = 1 - 4$). Stable and unstable regions of initial conditions in the (η, ρ_0) are shown [97]

8.5 Observation of Relativistic and Charge-Displacement Self-Channeling of Intense Subpicosecond Ultraviolet (248 nm) Radiation in Plasmas

The first experimental observation of the relativistic and charge-displacement self-channeling of an intense laser pulse is reported in [8]. For an electron density of $1.35 \times 10^{21}\,\text{cm}^{-3}$ and a power of the order of $3 \times 10^{11}\,\text{W}$, the results indicate a channel radius of less than $1\,\mu\text{m}$ and a maximal intensity $\sim 10^{19}\,\text{W/cm}^2$.

The experimental arrangement used in the study described in [8] is presented in Fig. 8.14a. A subpicosecond KrF* ($\lambda = 248\,\text{nm}$) laser [152] was used as the radiation source. It delivered linearly polarized radiation with a power of $\sim 3 \times 10^{11}\,\text{W}$ ($\sim 150\,\text{mJ}$, pulse duration $\sim 500\,\text{fs}$) in a beam with a diameter of $\sim 42\,\text{nm}$. When this radiation was focused into the chamber with lens LI (f/7), a focal radius $r_0 \sim 3.5\,\mu\text{m}$ was measured, giving a maximum intensity $I_0 \sim 8.6 \times 10^{17}\,\text{W/cm}^2$. The medium was provided by filling the chamber statically with gas [He, Ne, Ar, Kr, Xe, N_2, CO_2, or a mixture of Xe (4%) and N_2 (96%)] up to a maximum density of $\sim 1.89 \times 10^{20}\,\text{cm}^{-3}$.

The diffracted 248 nm radiation was measured as a function of the angle θ with respect to the direction of incident radiation propagation. The incident laser beam was blocked by a metal disk on the output window of the chamber, and lens L2 imaged the region near the focal zone on a fluorescent screen S. The diaphragm D in front of lens L2 restricted the diffracted light collection to a solid angle of $\sim 5°$ and simultaneously increased the field depth. The angle between the axis of the lens L2 and the axis of the incoming laser radiation could be readily varied up to a maximum angle of $\theta \sim 15°$. Two flat mirrors coated for high reflection ($\sim 99\%$) at 248 nm, both having a spectral bandwidth of $\sim 10\,\text{nm}$, served in reflection as spectral filters (F) for the diffracted laser radiation, so that only the scattered 248-nm radiation could illuminate the screen. An attenuator A was employed to adjust the intensity on the screen, and the images formed were recorded through the visible fluorescence produced with a microscope and a charge-coupled-device (CCD) camera.

The measured result for N_2 is shown in Fig. 8.14b. It corresponds to a density $\rho_{N_2} \sim 1.35 \times 10^{20}\,\text{cm}^{-3}$. To the left, a relatively large cone of light Rayleigh-scattered from the plasma is visible in the photographic inset at all angles, as the energy propagates toward the focal point of the lens, and, in the region to the right of the conical apex, a narrow filament developed. The diameter of this filament is not greater than $10\,\mu\text{m}$, which is the measured spatial resolution of the imaging system. The intensity distribution observed along the filament exhibited several bright features attributed to diffraction because they could not be seen for $\theta > 20°$. Since the axis of the imaging lens corresponded to an angle $\theta = 7.5°$, the scale along the abscissa of the

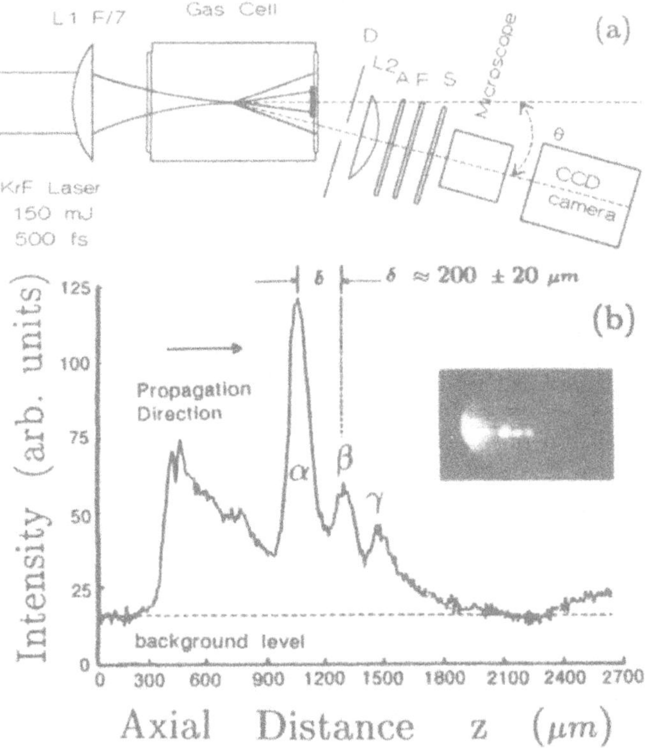

Fig. 8.14. (**a**) The experimental arrangement used in the study of relativistic and charge-displacement self-channeling. (**b**) Observation data for the propagation of a single intense laser pulse in N_2 at a density of $\sim 11.35 \times 10^{20}\,\mathrm{cm}^{-3}$. The maximum intensity is half the detector (CCD) saturation. The radiation is incident from the left. Inset: Photographic data with a vertical spatial resolution of $\sim 10\,\mu\mathrm{m}$. The graph illustrates the one-dimensional axial profile taken along the propagation direction (z) of the photographic data (inset). The spacing of the maxima, $\delta \simeq 200 \pm 20\,\mu\mathrm{m}$, is indicated [8]

photographic data is reduced by almost eightfold, giving a maximum filament length of $\sim 1\,\mu\mathrm{m}$. The graph in Fig. 8.14b represents the one-dimensional axial intensity distribution along the propagation direction (z) of the observed intensity pattern (inset). Three local maxima (α, β, γ) are visible with a spatial separation of $\delta = 200 \pm 20\,\mathrm{mm}$. The normal Rayleigh range for the focal geometry used was $\sim 200\,\mu\mathrm{m}$.

The filamentary channel diameter is an important quantity, that was estimated by measuring the maximum angular deviation of the diffracted light. The experimental value Φ of this diffracted cone was $\Phi \sim 20°$, a magnitude indicating a radius $r_\Phi \sim 0.9\,\mu\mathrm{m}$ through the relation $\Phi = 1.22\,\lambda/r_\Phi$. Filaments of this general nature were observed at densities above $\sim 1.35 \times 10^{20}\,\mathrm{cm}^{-3}$

in N_2, Ne, Ar, Kr, CO_2, and a mixture of Xe (4%) and N_2 (96%), but not in He and Xe, two materials discussed further below.

Two physical mechanisms of the medium's refractive index modification exist that can be responsible for the laser radiation propagation regime outlined above. They are (1) the Kerr effect stemming from ion shell polarization in the exterior field and (2) relativistic and charge-displacement nonlinearity discussed above in detail. An implication of the estimate of the channel radius ($r_\Phi \sim 0.9\,\mu m$) is that the laser radiation intensity in the above experiment is in the 10^{18}–$10^{19}\,W/cm^2$ range. According to the available experimental evidence on multiphoton ionization [153, 154] under these conditions He should be fully ionized and C, N, and 0 atoms constituting the molecular materials would retain, at most, only 1s electrons.

Consider explicitly the case of N_2, for which the estimated threshold intensities for producing N^{5+}, N^{6+}, and N^{7+} are $1.6 \times 10^{16}\,W/cm^2$, $6.4 \times 10^{18}\,W/cm^2$, and $1.3 \times 10^{19}\,W/cm^2$, respectively [153, 154]. Hence, the volume of the channel would be largely ionized to N^{5+}, and certain localized high-intensity regions contribute some N^{6+}. The consequences of this ionization pattern are that (1) the contribution from the Kerr effect is small since the polarizabilities of the remaining 1s electrons are low, and (2) the electron concentration n_e initially produced in the focal region is nearly uniform which means that $n_e \approx 1.35 \times 10^{21}\,W/cm^2$ for the data on N_2 shown in Fig. 8.14b.

The relativistic and charge-displacement critical power P_{cr} defined by (8.11) can be rewritten in terms of the plasma electron concentration as [8]

$$P_{cr} \approx 1.62 \times 10^{10}(n_{cr}/n_e)\,W\,, \tag{8.28}$$

where n_{cr} is the critical electron density (for $\lambda = 248\,nm$, $n_{cr} = 1.82 \times 10^{22}\,cm^{-3}$).

The critical powers for the experimental conditions, for He and N_2 at a medium density $\rho = 1.35 \times 10^{20}\,cm^{-3}$, are $1.08 \times 10^{12}\,W$ and $2.19 \times 10^{11}\,W$, respectively. Therefore, since the incident power was $P \approx 3 \times 10^{11}\,W$, no filament could emerge in He, and the above experimental finding agrees with this prediction. Moreover, the diffracted cone of radiation was also absent in the experiment with He. In contrast, for N_2 $P/P_{cr} \approx 1.37$, a condition that held generally ($P/P_{cr} > 1$) for all materials that exhibited evidence of channel formation. Some contribution from the Kerr effect may be present, even for light materials (Ne, N_2, and CO_2) in the early stage of channel formation prior to the development of a substantial level of ionization. The Kerr effect may play a greater role in heavier gases (Ar, Kr, and Xe). A specific estimate of the nonlinear index change arising from both N^{5+} and N^{6+} at an intensity of $10^{19}\,W/cm^2$ indicates that their contribution is less than 10^{-3} that of free electrons; hence the ionic contribution can be neglected in N_2 for the conditions studied.

Paper [8] included a comparison between the theoretical results presented above and the experimental findings for N_2. Both the longitudinal intensity profile and the radial extent of the channel were compared. Figure 8.15 illustrates the intensity profile $I(r,z)/I_0$ calculated for the physical parameters corresponding to those of the experiment (i.e., $P \approx 3 \times 10^{11}$ W, $r_0 = 3.5\,\mu$m, $n_e \approx 1.35 \times 10^{21}$ cm^{-3}, and $P/P_{cr} > 1$). Importantly, all of these parameters are obtained from independent measurements of (1) the laser pulse parameters (including determinations of the energy and power P), (2) the incident radiation focal radius r_0, and (3) the characteristics of the multiphoton ionization [154], that generated electron concentration n_e. The calculated normalized electron concentration is presented in Fig. 8.15, from which it is seen that electronic cavitation occurs only near the positions of intensity profile maxima. The curve in Fig. 8.16 is the one-dimensional axial intensity distribution $I(0,z)/I_0$ for Fig. 8.15. The theoretically predicted spacing δ of the maxima is $\delta \sim 185\,\mu$m, a value in close agreement with the experimental data ($\delta \sim 200 \pm 20\,\mu$m) shown in Fig. 8.14b.

The measurements indicated an approximate value of $r_\Phi \sim 0.9\,\mu$m for the radial extent of the channel, a result that can be compared with the corresponding theoretical figure. Five radial intensity distribution cross sections $I(r,z_i)/I_0$ for the waveform pictured in Fig. 8.15 are shown in Fig. 8.16. Since the measurement of this angularly scattered radiation did not correspond to a known longitudinal position, this comparison can be only qualitative, but the radial distributions shown indicate that the expected value lies in the interval $0.5 \leq r \leq 1.0\,\mu$m, a range that comfortably includes the experimental value r_Φ.

The results observed with Xe require additional explanations since those experiments did not give evidence of channel formation. In significant contrast to the N_2 case, the electron concentration n_e, produced by multiphoton ionization [154] in Xe is expected to be very nonuniform spatially. For intensities spanning 10^{16}–10^{18} W/cm^2, the corresponding density n_e would vary by more than a factor of 2. Since this nonuniformity would tend to reduce the refractive index locally in the central high-intensity region, significant defocusing action is expected which could suppress channel formation.

In a different study described in [99], an observation of channels spanning 3–4 mm was reported. Significantly, this distance spans over 100 Rayleigh ranges. In this experiment, optimal conditions were established by matching the incident laser beam parameters close to those given by the lowest eigenmode of the Schroedinger equation with relativistic and charge-displacement nonlinearity and, as a result, in contrast to the case described above, the channels observed were longitudinally uniform.

In this study, a 270-fs, 248-nm laser pulse with a beam diameter of approximately 70 mm and power of approximately 460 GW was focused into a differentially pumped gas cell. The measured focal zone was $3\,\mu$m, and the incident maximal focal intensity was approximately 7×10^{18} W/cm^2. The gas pressure in the target chamber was varied from 760 to 3800 Torr. In particular, a single-shot, time-integrated picture of a filament produced by a

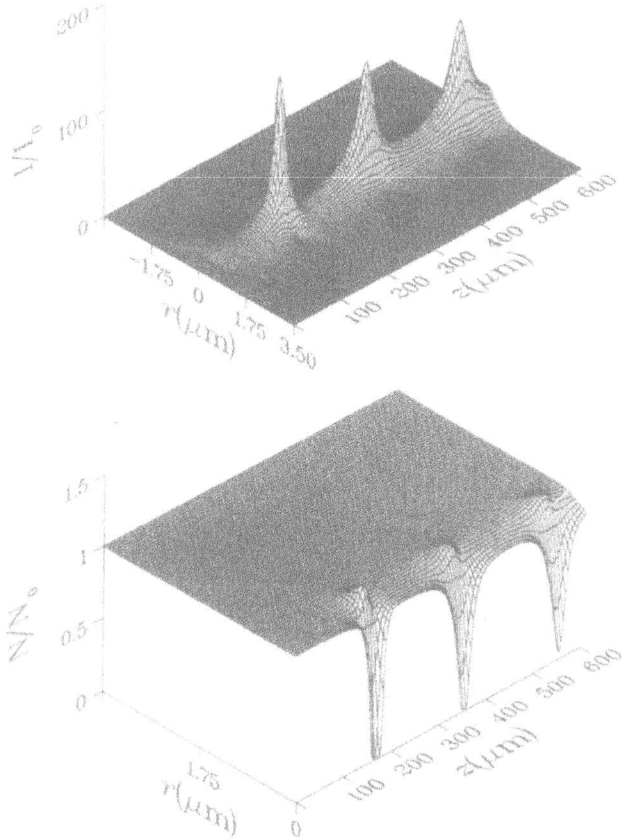

Fig. 8.15. Simulation results for N_2 with $P = 3 \times 10^{11}$ W, $r_0 = 3.5\,\mu m$, $n_e \sim 11.35 \times 10^{20}\,cm^{-3}$, $I_0 = 8.6 \times 10^{17}\,W/cm^2$. Normalized intensity $I(r,z)/I_0$ and normalized electron concentration $n(r,z)/n_0$ [8]

laser pulse with the above parameters in Kr at a pressure of 3870 Torr was recorded. Digitized output corresponding to the axial line-out of this image is shown in Fig. 8.17. Due to the details of the experimental configuration used, no light is visible before the axial location of the laser focus,. Thus, there is no signal at distances less than 1800 µm. As one can see in the above figure, the length of the laser channel exceeds 3 mm, a quantity greater than 100 Rayleigh ranges. The estimated channel diameter was 1.4 µm. A gradual intensity attenuation at large propagation distances ($z > 420\,\mu m$) results in termination of self-channeling when the beam power becomes lower than the relativistic and charge-displacement critical power.

The emission of X rays from the laser beam propagation zone provides additional evidence of channeled propagation. This emission is attributed to bremsstrahlung by energetic electrons inside the channel where the energy density is extremely high. An X-ray detector was used in the framework of the

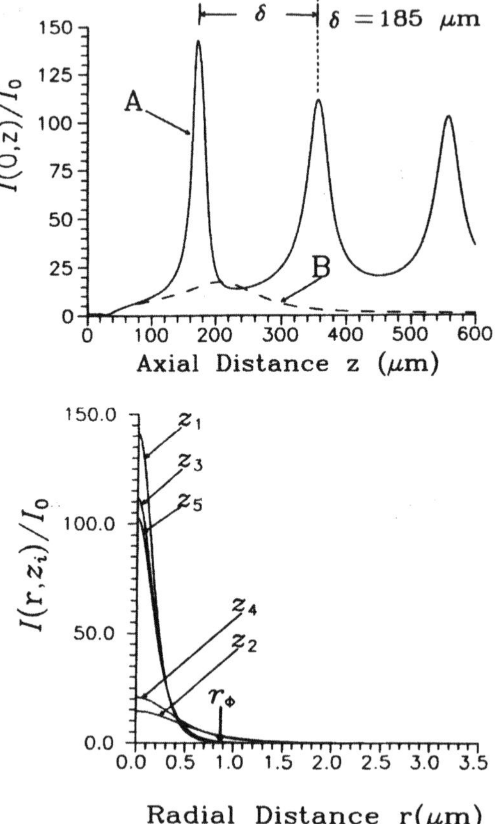

Fig. 8.16. One-dimensional intensity distributions from the plots shown in Fig. 8.15. Curve A-normalized one-dimensional intensity distribution $I(0,z)/I_0$ on the propagation axis, $\delta = 185\,\mu\text{m}$ (curve B, calculation with charge-displacement term neglected) and normalized radial intensity profiles $I(r,z_i)/I_0$. Longitudinal positions $z_1 = 172\,\mu\text{m}$, $z_2 = 245\,\mu\text{m}$, $z_3 = 358\,\mu\text{m}$, $z_4 = 441\,\mu\text{m}$, $z_5 = 559\,\mu\text{m}$, and $r_\Phi = 0.9\,\mu\text{m}$ [8]

study described in [99] to determine the correlation between X-ray production and channel formation. A cylindrically shaped detector was mounted on one of the target cell ports with a window located at a distance of 0.5–1 mm from the self-channeling zone to view it perpendicularly to the laser beam propagation direction. The window that consisted of a 10-μm thick Be foil and permitted the transmission of radiation with a quantum energy beyond 0.5 keV was placed over the conical end of the detector which had a 1-mm diameter aperture. A sodium salicylate scintillator was approximately 3 mm behind the aperture. The sodium salicylate was dissolved in methanol and then applied to the end of a fiber-optic rod that extended along the detector axis and through a vacuum flange. To reduce the X-ray absorption between

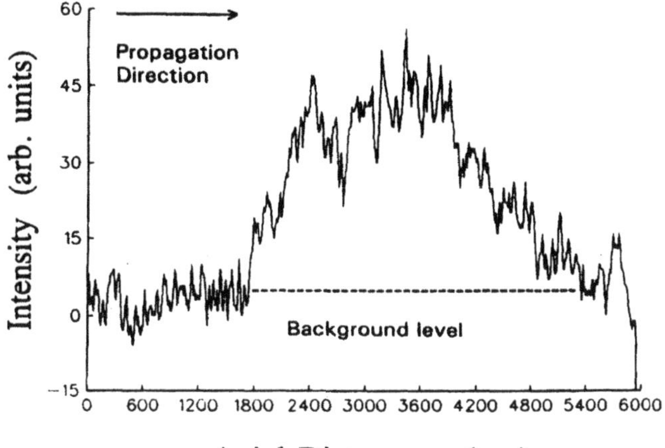

Fig. 8.17. Single-shot, time-integrated filament obtained from the channeled propagation of a 460 GW laser pulse in Kr at a pressure of 3870 Torr. Digital readout corresponding to an axial line-out of the experimental data indicating laser beam self-channeling. The measured channel length was greater than 3 mm, which is more than 100 Rayleigh ranges [99]

the window and the scintillator, the system was evacuated to a pressure of several millitorr. A photomultiplier was attached to the other end of the fiber-optic rod, and the photomultiplier signal was recorded by an oscilloscope.

The dependence of the X-ray outcome on the target gas pressure is depicted in Fig. 8.18. The X-ray emission in N_2 was recorded at pressures in the 760–1000 Torr range. Simultaneous observations of the diffracted light were made to establish that self-channeling occurs at the same pressure. The conclusion is that X-ray emission is registered only simultaneously with relativistic and charge-displacement self-channeling.

The stability map used to optimize the experimental conditions in [99] is presented in Fig. 8.19. In this map, point A corresponds to the experimental conditions reported in [8] ($r_0 = 3.5\,\mu\text{m}$, $P_0 = 0.3\,\text{TW}$, $N_{e,0} = 1.35 \times 10^{21}\,\text{cm}^{-3}$, which is equivalent to $\rho_0 = 23$ and $\eta = 1.37$), point B to the experimental parameters corresponding to the data illustrated in Fig. 8.17 ($r_0 = 1.5\,\mu\text{m}$, $P_0 = 0.5\,\text{TW}$, $N_{e,0} = 1.4 \times 10^{21}\text{cm}^{-3}$, or to $\rho_0 = 10$ and $\eta = 2.36$), and point C to the range of conditions under which an X-ray yield is observed. Obviously, a greater self-channeling distance as well as a more uniform channel longitudinal structure are obtained by matching the incident laser beam parameters closer to those given by the lowest eigenmode of the Schroedinger equation with relativistic and charge-displacement nonlinearity.

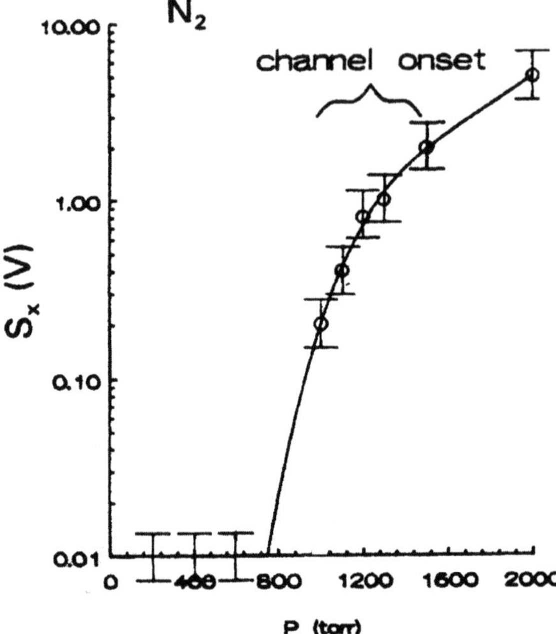

Fig. 8.18. Pressure dependence of the X-ray signal strength observed in N_2. Correlation of the X-ray signal with the observation of channel formation, which was observed on a single-shot basis, allows X-ray production to serve as a diagnostic for laser beam self-channeling [99]

8.6 Conclusions

A theoretical approach for investigating the two-dimensional dynamics of propagation of coherent, ultrashort ($\tau_{\rm ion} \gg \tau \gg \tau_{\rm e}$) relativistically intense laser pulses in cold underdense plasmas has been described. Four basic physical phenomena are included within the model used: (1) the refractive index nonlinearity due to the relativistic increase in the mass of plasma free electrons, (2) the refractive index variation resulting from the plasma electron concentration modification by the ponderomotive force, (3) the diffraction of laser radiation, and (4) the laser radiation refraction caused by transversely inhomogeneous electron concentration distributions.

The main conclusions are the following [7,8,97,99]:

- The cooperative effect of relativistic and charge-displacement nonlinearities leads asymptotically to stable high-intensity, z-independent modes of self-channeling, and a major fraction of the incident laser beam power can be confined in these paraxial modes. Stable cavitation of the plasma's electron component is typical for these spatially confined modes.
- The z-independent modes, to which the solutions of the equation describing relativistic and charge-displacement propagation tend asymptotically, are

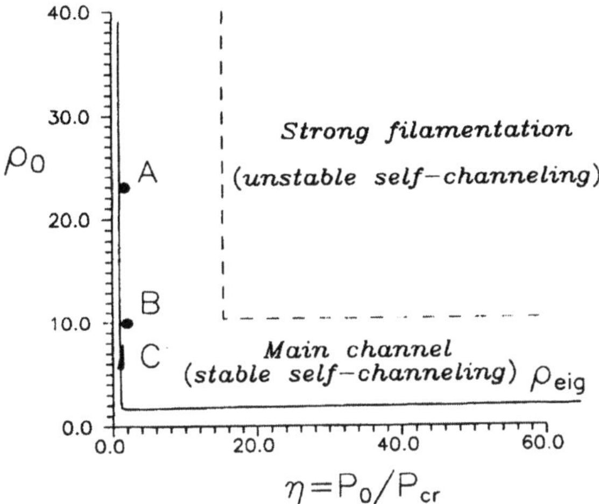

Fig. 8.19. Stability map for relativistic and charge-displacement self-channeling. Point A represents the experimental conditions of [8] (see previous section). Point B represents the experimental conditions corresponding to the data illustrated in Fig. 8.17. Point C represents the experimental conditions that resulted in the observation of X rays, as illustrated in Fig. 8.18. Optimization of the experimental parameters by minimizing ρ_0 for a given value of η makes it possible to achieve stable self-channeling [99]

recognized as the axially symmetrical *lowest eigenmodes* of the governing nonlinear Schroedinger equation.

- In the case of small azimuthal perturbations, maps representing the filamentation stability criteria are developed in the plane of the two variables $\rho_0 = r_0\omega_{\mathrm{p},0}/c$ and $\eta = P_0/P_{\mathrm{cr}}$. Stable zones span extensive areas of physically attainable parameters. The stability of the lowest axially symmetrical relativistic and charge-displacement eigenmodes is particularly robust.
- Experiments examining a new relativistic regime of high-intensity pulse propagation in plasmas have been performed, and the findings indicate the formation of a channeled propagation mode over a length considerably greater than the Rayleigh range. Comparisons of the experimental observations with the theoretical predictions produce excellent agreement for both the longitudinal structure of the intensity profile and the radial extent of the channel.
- Under optimal conditions, namely, when the incident relativistically intense laser beam parameters are matched to the lowest eigenmode of the Schroedinger equation with relativistic and charge-displacement nonlinearity, the channeled propagation spans a distance greater than 100 Rayleigh ranges.

9. Dynamics of Relativistic and Charge-Displacement Self-Channeling in Time and 2D Space

The topic of this chapter is a description of the superintense laser beam temporal dynamics in media with relativistic and charge displacement nonlinearity in the framework of the nonlinear Schroedinger equation model and self-channeling simulations on the basis of the more advanced wave equation model. The former approach makes it possible to establish a picture of laser pulse dynamic self-modulation in a plasma and the giant broadening of its spectra entailed by this effect, as well as to examine the dynamics of the two-dimensional intense laser pulse solitons. The advantage of the latter one is the complete description of the wave properties of propagation. In particular, simulations performed with the help of the wave equation with relativistic and charge-displacement nonlinearity illustrate the process of the prepulse formation. The numerical technique for treating the wave equation studied is presented in [161], and the numerical results are discussed below following [115]. The simulations are carried out in the axially symmetrical case.

9.1 Superintense Two-Dimensional Solitons, Self-Modulation, and Spectral Broadening

As the first step, let us examine a number of propagation simulation results obtained in the framework of the modified nonlinear Schroedinger equation model. These results make it possible to illustrate several aspects of the general propagation character at relativistic intensities that remained outside the scope of the studies described in the previous chapter.

9.1.1 Laser Beam Stabilization and the Formation of a Two-Dimensional Solitary Wave

Consider the interaction of an intense laser beam with a cold underdense plasma, where the propagation is governed by relativistic and charge-displacement nonlinearity. It is shown in [101] that in this case, the asymptotic limit of the laser radiation amplitude distribution is a two-dimensional soliton. Below, we follow the study performed in this work.

There are no derivatives with respect to the variable η in (8.12). Therefore, in the framework of the approach to this equation, the laser pulse is treated as consisting of transverse "slices", and the propagation of each of them can be simulated independently. This is the method implemented in the previous chapter and the references therein. As we have seen, whenever the condition $P_0 > P_{\mathrm{cr}}$ is met for a particular slice (assuming an initially flat phase front), the corresponding solution of the Schroedinger equation with relativistic and charge-displacement nonlinearity tends asymptotically at large z to this equation's lowest eigenmode. According to the results described in the previous chapter, spatially localized eigenmodes exist for $0 < s < 1$. The parameter s of the eigenmodes emerging dynamically at $z \to \infty$ for a particular slice depends on the initial conditions. Since those depend implicitly on $\eta = t - v_{\mathrm{g}}^{-1} z$, generally $s = s(\eta)$. Consequently, for the asymptotic state of the entire laser pulse, one can write

$$a(z \to \infty, r, \eta) = U_{s(\eta),0}(r) \exp\{\mathrm{i}\,[s(\eta) - 1]\,\eta\} \tag{9.1}$$

and

$$0 < s(\eta) < 1\,. \tag{9.2}$$

The latter condition means (at least for laser pulses with initially flat wave fronts and Gaussian or hyper-Gaussian initial transverse intensity distributions) that only the temporal parts for which $P_0 > P_{\mathrm{cr}}$ contribute to the asymptotic state, whereas the remaining ones are diffracted away at the periphery.

In the general case, the asymptotic solution defined by (9.1) represents a two-dimensional soliton. It is also possible that the laser pulse undergoes disintegration, and the result is the formation of several two-dimensional solitons; each of them carries a power greater than the critical power of relativistic and charge-displacement self-channeling. Then, the sum of the soliton powers is less than the initial laser beam power.

Below, the transformation of an intense laser pulse into a two-dimensional soliton is illustrated by a specific example presented in [101]. A solution of (8.12)–(8.14) corresponding to excimer laser radiation with wavelength $\lambda = 248\,\mu\mathrm{m}$, initial aperture $r_0 = 3\,\mu\mathrm{m}$, duration $\tau_0 = 30\,\mathrm{fs}$, and maximal initial intensity $I_{\mathrm{m}} = 4.5 \times 10^{18}\,\mathrm{W/cm^2}$ is shown in Fig. 9.1. It is assumed that the initial temporal and radial profile of the laser pulse is Gaussian. This figure illustrates a sequence of laser pulse snapshots at 58.1, 86.2, 91, 103, 121.7, and 153.6 fs which demonstrates clearly how the laser pulse propagating in a plasma evolves into a two-dimensional soliton. The following details of this process can be noted. In the beginning, a powerful focus is formed due to relativistic and charge-displacement self-focusing. Ponderomotive charge displacement plays a substantial role in the focal area as an additional mechanism of laser radiation nonlinear self-trapping. Subsequently, the focus splits into two maxima moving in the opposite directions within the laser pulse envelope. The result is a strong longitudinal modulation of the laser beam.

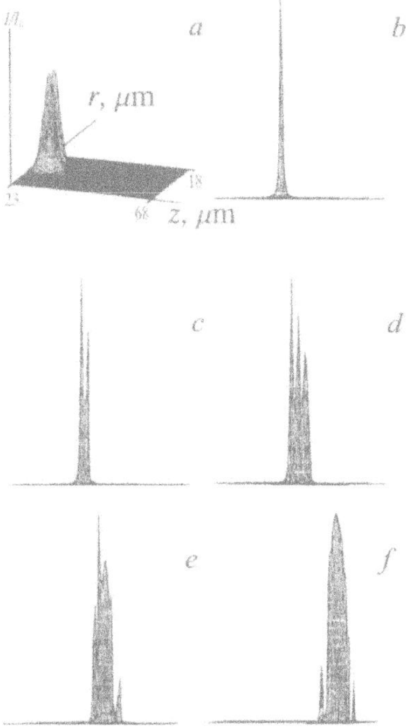

Fig. 9.1. The propagation of a laser pulse with an initially Gaussian temporal profile and an initially plateaulike radial intensity distribution in a semi-infinite plasma under the influence of relativistic and charge-displacement nonlinearity. Snapshots correspond to times (**a**) 58.1, (**b**) 86.2, (**c**) 91, (**d**) 103, (**e**) 121.7, and (**f**) 153.6 fs. The ratios of the maximal laser beam intensity and the maximal initial intensity at the above times are (**a**) 7.6, (**b**) 497.7, (**c**) 497.7, (**d**) 217.0, (**e**) 243.9, (**f**) 193.4 [101]

The pulse is transformed into a number of overlapping local maxima. The modulation decreases gradually at the laser pulse central part, and the intensity is stabilized. The modulation at the laser pulse periphery takes longer to stabilize. A spatial domain with a lower plasma electron concentration is formed within the soliton.

Noted that extremely high power is concentrated in the two-dimensional soliton described in this section. For this reason, this waveform is of great interest for a number of prospective applications.

9.1.2 Giant Broadening of Laser Pulse Spectra

There is a number of studies of the spectra of picosecond laser pulses propagating in nonlinear media [148–150]. It is well-known that the broadening of these spectra can be due to self-modulation of laser radiation, or to the character of laser pulse phase evolution. Pulse modulation causes a beam

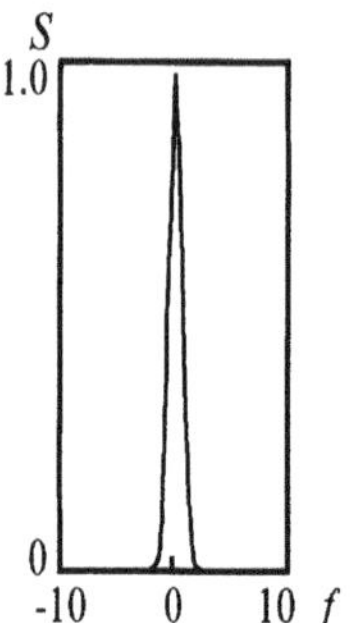

Fig. 9.2. The spectrum of laser radiation incident on a plasma

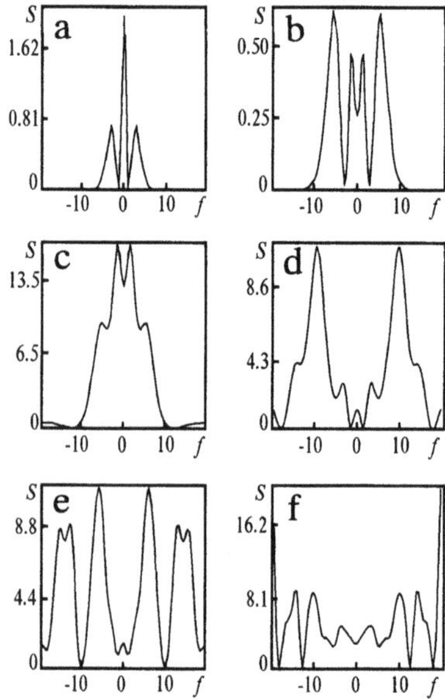

Fig. 9.3. Laser pulse spectrum at a sequence of points on the propagation axis corresponding to the intensity evolution depicted in Fig. 9.1 [100]

to disintegrate into a series of narrow maxima. The phase evolution results in shifting the laser pulse frequency. The model based on the Schroedinger equation with relativistic and charge-displacement nonlinearity describes the former of the above two effects. Since the dispersion of the propagation of different laser pulse slices is neglected in the framework of this model, it does not describe the laser pulse spectra broadening owing to phase evolution.

The laser pulse temporal spectra at a sequence of points on the propagation axis given by $z_k = 29.03$, 38.57, 40.43, 47.27, 53.54, and 57.53 μm are calculated in [100,151] (an initially plateaulike laser intensity distribution and relativistic and charge-displacement nonlinearity are considered). The input signal spectrum is shown in Fig. 9.2. The dynamic frequency spectra are depicted in Fig. 9.3. These spectra are normalized by the maximal magnitude of the input signal frequency spectrum:

$$S(\omega, z_k) = \frac{|F(\omega, z_k)|^2}{|F(0,0)|^2}, \qquad F(\omega, z_k) = \frac{1}{2\pi} \int_{-\infty}^{\infty} u(t, z_k)\, e^{-\mathrm{i}\omega t}\, \mathrm{d}t\,. \tag{9.3}$$

Here, $\omega = \pi f / \tau_c$ is the frequency (Hz) and f stands for normalized frequency. Note that the spectra are symmetrical with respect to the central frequency due to the symmetry of the laser pulse's temporal profile.

The important feature of the laser pulse spectra illustrated by Fig. 9.3 is their giant broadening compared to the initial profile. The spectra broaden by a factor of 15–20 or even more. Spectral broadening is a manifestation of transformation of the laser pulse envelope into a sequence of narrow local maxima. Their duration is inversely proportional to the width of the laser pulse spectrum.

9.2 Nonlinear Wave Equation Model

It is wellknown that the propagation of electromagnetic waves in a vacuum is described by a linear wave equation which is derived from the Maxwell equations. At low intensities, the propagation of light in matter is also governed by a wave equation with constant coefficients (linear optics). The transport of intense electromagnetic radiation in matter is often described by a nonlinear wave equation with intensity-dependent coefficients. Typically, additional approximations are used to treat such wave equations, and we have seen examples of this approach in previous chapters.

The most common method of dealing with laser pulse propagation is reduction of the wave equation model to the nonlinear Schroedinger equation for slow amplitude of the electromagnetic field (of the see Chap. 6). This is possible when the amplitude of the electromagnetic field does not exhibit substantial variations on the radiation wavelength scale in the propagation direction and on the optical field's oscillation period timescale. When these criteria are met throughout the propagation distance, the corresponding nonlinear Schroedinger equation remains applicable, though a quantitative estimate of the difference between the solutions of the nonlinear wave and Schroedinger equations is an interesting issue, nonetheless. However, the nonlinear Schroedinger equation model is often used to simulate stationary and also short and ultrashort ($\tau < 1\,\mathrm{ps}$) laser pulse propagation. It is wellknown that in such cases nonstationary waveforms emerge comprising powerful moving foci with sharp variations of the optical field's amplitude. The Schroedinger equation represents a less reliable model in focal areas. Besides, under certain conditions, nonlinear wave equation's solutions can be unstable, and their small perturbations with certain periods grow in time.

Consequently, the small deviations of solutions of the nonlinear Schroedinger equation from those of the general wave model which "accumulate" in the first focal area might evolve into substantial quantities. Therefore, the applicability of the nonlinear Schroedinger equation after the emergence of the first focus is an issue that requires careful examination.

Note additionally that there are models which can be considered intermediate between the nonlinear Schroedinger and wave equations. These models include some, but not all, of the second derivatives in co-moving variables, that are neglected when the Schroedinger equation is derived from the wave one. Models of this type called modified nonlinear Schroedinger equations, describe the higher order laser radiation dispersion.

Formulation of the Nonlinear Wave Equation Model. Following [12, 115], we present the nonlinear wave equation describing relativistic and charge-displacement self-channeling in the form

$$\left(\frac{1}{v_g}a_t + \partial_z a\right) + \frac{\mathrm{i}}{2k}\left[(\Delta_r + \partial_z^2 - c^{-2}\partial_t^2) + k_p^2\left(1 - \frac{1 + k_p^{-2}\Delta\gamma}{\gamma}\right)\right] a = 0\,. \tag{9.4}$$

Obviously, it differs from (6.6) and (6.16) by an alternative normalization of variables. The initial and boundary conditions for this equation are

$$a|_{t=0} = a_0(r,z)\,, \tag{9.5}$$

$$(v_{\mathrm{g}}^{-1}\partial_t + \partial_z)\, a|_{t=0} = f(r,z)\,, \tag{9.6}$$

$$\partial_r a|_{r=0} = 0\,, \tag{9.7}$$

$$a|_{z=\pm\infty} = 0\,, \qquad a|_{r=+\infty} = 0\,. \tag{9.8}$$

In terms of real-valued functions, (9.5) includes the initial conditions for laser radiation intensity and phase.

If laser radiation propagates in an unbounded medium ($-\infty < z < \infty$) and the laser pulse shape at $t = 0$ is a known function, it is convenient to introduce the variables $\eta = v_g t - z$ and $\tau = t$ to describe spatially localized waveforms. In this case, (9.4) becomes

$$\frac{1}{v_{\mathrm{g}}}\partial_\tau a + \frac{\mathrm{i}}{2k}\left[\Delta_\perp + \left(1 - \frac{v_{\mathrm{g}}^2}{c^2}\right)\partial_\eta^2 - \frac{2v_{\mathrm{g}}}{c^2}\,\partial_{\eta\tau}^2 - \frac{1}{c^2}\,\partial_\tau^2\right] a + k_{\mathrm{p}}^2\left[1 - \frac{1 + k_{\mathrm{p}}^{-2}\left(\Delta_\perp\gamma + \partial_\eta^2\gamma\right)}{\gamma}\right] a = 0\,. \tag{9.9}$$

In co-moving variables, the above initial and boundary conditions are written as

$$a(r,\eta,\tau = 0) = a_0(\eta, r)\,, \tag{9.10}$$

$$\partial_\tau a(r, \eta, \tau = 0) = f(\eta, r)\,, \tag{9.11}$$

$$\partial_r a(r = 0, \eta, \tau) = 0\,, \tag{9.12}$$

$$a(r = \infty, \eta, \tau) = 0\,, \qquad a(r, \eta = \pm\infty, \tau) = 0\,. \tag{9.13}$$

In the envelope approximation where the complex field amplitude is a slow function on the laser radiation wavelength spatial scale and on the field oscillation period timescale, $|\partial_z a|$, $c^{-1}|\partial_t a| \ll k|a|$. In this case, a typical approach is to use the nonlinear Schroedinger equation instead of (9.9).

Due to the complexity of the problem involving the nonlinear wave equation, it is natural to consider several models that are intermediate between the nonlinear Schroedinger and the wave equations. Generally six problems of this type exist. First, this refers to the nonlinear Schroedinger equation retaining one of the following second derivatives: ∂^2_η, $\partial^2_{\eta\tau}$, and $\partial^2_{\tau\tau}$. Secondly, a nonlinear Schroedinger equation can include a pair of any two of the above derivatives. The corresponding equations are called modified nonlinear Schroedinger equations.

9.2.1 A Comparison of Simulations Based on the Modified Nonlinear Schroedinger Equation and on the Nonlinear Wave Equation

Assume that the laser pulse intensity distribution is a hyper-Gaussian function of the radial and longitudinal coordinates at $\tau = 0$:

$$|a_0(\eta, r)|^2 = a_0^2 \exp\Big\{-\Big[(\eta - \eta_1)/\eta_0\Big]^{N_1} - (r/r_0)^{N_2}\Big\}\,, \tag{9.14}$$

where $\eta_0 = \tau_0 v_g$ and η_1 is the intensity maximum coordinate. Let the pulse have a uniform plane phase front:

$$a_0(\eta, r) = |a_0(\eta, r)| \exp\left[i\varphi_0(\eta, r)\right]\,, \qquad \varphi_0(\eta, r) = \text{const}\,. \tag{9.15}$$

We also use the condition,

$$f(\eta, r) = 0\,,$$

for the second derivative, which, in terms of the linear optics, would have meant preserving the shape of the propagating laser pulse.

Problem (9.9)–(9.13) was solved in [161] with the help of a spectral finite-difference method in an unbounded domain given by $r > 0$ and $-\infty < \eta < \infty$ [162]. It can be considered in two semi-infinite domains in η, namely, one spanning from the laser pulse front at $\eta = 0$ to $\eta = \infty$ and one from the pulse front to $\eta = -\infty$. The numerical technique involves the collocation method, and the solution is found as a finite sum in the Chebyshev polynomials of the first kind. In each domain, the method leads to a set of second-order ordinary differential equations in time. Continuity of the solution and its derivative at $\eta = 0$ are additional conditions for solving the above equations. In general, the

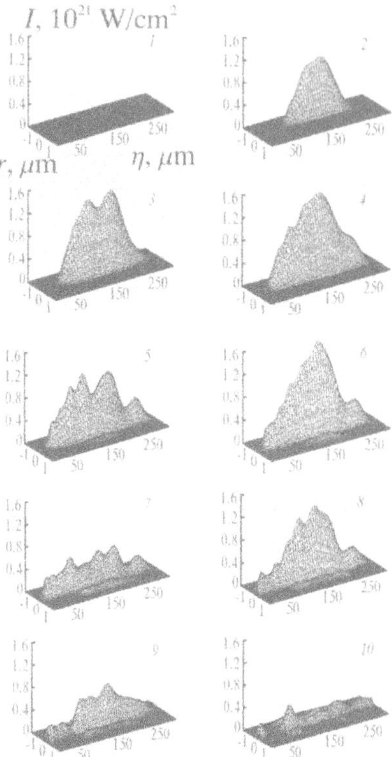

Fig. 9.4. The propagation of a laser pulse with an initially Gaussian temporal and radial profile in an unbounded plasma simulated using the wave equation with relativistic and charge-displacement nonlinearity. The maximal beam power is given by $P_0/P_{cr} = 22.3$ [115]

present technique can be regarded as a version of the finite element method. The nodes of the collocation grid are located more densely near the $\eta = 0$ point on both sides of it to provide for an adequate simulation of the prepulse formation. The ordinary differential equations are solved using the predictor–corrector scheme where the second-order precision scheme is the predictor.

Consider the propagation of an initially Gaussian excimer laser pulse ($N_1 = 2$, $N_2 = 2$) with the following parameters: wavelength $\lambda = 248\,\mu$m, aperture $r_0 = 3\,\mu$m, and energy $E = 6$ J. The plasma electron concentration is assumed to be $n_e = 7.5 \times 10^{20}\,\text{cm}^{-3}$. Figure 9.4 shows the corresponding solutions of the nonlinear wave equation with relativistic and charge-displacement nonlinearity at the following times: $\tau = 93.75 + 18.75 \times j$ fs, where $j = 1, 2, \ldots, 10$.

The initial Gaussian intensity distribution is not depicted in Fig. 9.4. The magnitude of its intensity is over a 100 times lower, and its transverse aperture is 25 times greater than those of the distributions shown. The intense laser

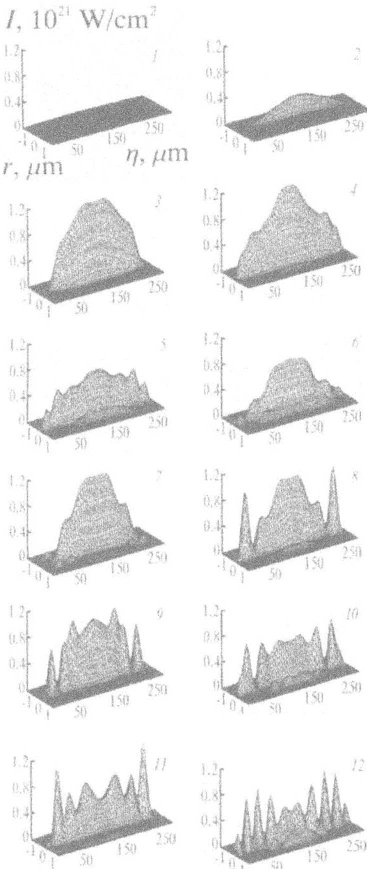

Fig. 9.5. The propagation of a laser pulse with an initially Gaussian temporal and radial profile in an unbounded plasma simulated using the modified Schroedinger equation with relativistic and charge-displacement nonlinearity. The maximal beam power is given by $P_0/P_{\mathrm{cr}} = 22.3$ [115]

pulse transverse aperture collapses at the axis of propagation. The first and second foci of the pulse are illustrated in Fig. 9.4.

To analyze the applicability of simpler models to the situation discussed, we present a solution of the modified nonlinear Schroedinger equation with the term ∂_η^2 in Fig. 9.5. The pertinent physical conditions are the same as above. The waveform snapshots in Fig. 9.5 correspond to $\tau = 37.5\,\mathrm{fs}$ and $\tau = 37.5 \times (j+1)\,\mathrm{fs}$, where $j = 2, \ldots, 12$.

The only difference in posing the problem for the modified nonlinear Schroedinger equation from that for the wave equation is that in this case there is no initial condition for the first derivative.

Note that the solutions of the nonlinear Schroedinger equation and its modification involving the term $a_{\eta\eta}$ are compared in [103]. The basic conclusion drawn in that paper is that differences between the two solutions accumulate at large propagation distances. Besides, it is important that the solutions to the nonlinear wave equation are not symmetrical in η. This circumstance is due to the presence of the $a_{\eta\tau}$ term in the "complete" model. As for the solutions of the modified Schroedinger equation, obviously they are symmetrical in η. Consequently, the impact of $a_{\eta\tau}$ on the overall solution character is significant.

A quantitative proximity in the evolution of the foci predicted by the nonlinear wave equation and the modified Schroedinger equation models can be found only at the initial stage of the emergence of the first focus. Further on, the solution of the nonlinear wave equation differs substantially from that of the Schroedinger equation. On the qualitative level, the distinction is mainly due to the influence of the $c^{-2}a_{\tau\tau}$ term. The laser pulse self-modulation in the η coordinate begins earlier than predicted by the modified nonlinear Schroedinger equation, which follows from the comparison between Fig. 9.4(3) and Fig. 9.5(3). Therefore, the second derivative in time plays an important role as well.

A comparison of Fig. 9.4 and Fig. 9.5 demonstrates that, according to the predictions of the wave equation model, a fraction of the laser radiation propagates forward ahead of most of the laser pulse, i.e. a prepulse is formed. Note that smaller values of η in the figures correspond to the laser pulse front. The nonlinear wave equation model predicts stretching of the laser pulse along the propagation axis, an effect seen clearly in Fig. 9.4.

As we have seen, neither of the derivatives $\partial^2_{\eta\eta}$, $\partial^2_{\eta\tau}$, or $\partial^2_{\tau\tau}$ can be neglected even for picosecond laser pulses. The problem becomes even more complicated when yet shorter laser pulses are considered.

A question naturally arising in this context is, What is the place of the nonlinear Schroedinger equation model and its modifications in the physics of ultrashort laser pulses? From our point of view, the situation is as follows. The models based on the nonlinear wave equation require massive computations, whereas the solution process for the Schroedinger equation is substantially less costly. The nonlinear Schroedinger equation describes laser pulse propagation adequately at relatively short propagation distances, specifically, at distances prior to the formation of the first focus and somewhat past it. The Schroedinger equation approximation performs better for pulses of greater duration ($\tau \gg 10^{-11}$ s). In ultrashort pulse physics, the results obtained with the help of the nonlinear Schroedinger equation should be interpreted mostly on the qualitative level, but they still indicate the overall tendencies of the propagation character, for example, beam transverse aperture collapse, nonlinear beam self-trapping, dynamic self-modulation. However, the nonlinear wave model is necessary for precise modeling.

9.2.2 Laser Pulse Self-Modulation in a Self-Channeling Regime

An interesting effect which also can be described in the framework of the nonlinear wave equation model of laser pulse propagation is spatial and temporal self-modulation. In this chapter, we examine the nonlinear stage of this process.

Simulation Results. Consider the (t, r, z) dynamics of relativistically intense laser radiation in plasmas. In contrast to the results discussed in the previous section, here our attention will be focused on the near-threshold propagation regime. In the case studied in the previous section, the ratio of the initial beam power and the critical power of relativistic and charge-displacement self-channeling is $P_0/P_{\mathrm{cr}} \approx 22.3$. In what follows, this value is much lower, specifically, $P_0/P_{\mathrm{cr}} \approx 1.4$.

The three-dimensional solutions of the wave equation with relativistic and charge-displacement nonlinearity are visualized as a sequence of the laser pulse intensity distribution snapshots. Fig. 9.6 illustrates a solution of the wave equation with relativistic and charge-displacement nonlinearity at times given by $t = 1000 + 31.25 \times j\,\mathrm{fs}$, where $j = 1, 2, \ldots, 20$. The simulation parameters are $\lambda = 248\,\mu\mathrm{m}$, $I_0/I_{\mathrm{r}} = 0.019$, $r_0 = 3.5\,\mu\mathrm{m}$, and $\tau_0 = 400\,\mathrm{fs}$. Here, I_0 is the maximal initial intensity. The solutions are depicted in co-moving variables $\eta = v_{\mathrm{g}}t - z$ and $\tau = t$, and the laser pulse front corresponds to $\eta = 0$.

The following features of the laser pulse nonlinear evolution are revealed by the simulation. A central maximum is formed at the time corresponding to $j = 1$. Later $(j = 2 - 4)$ two more maxima emerge; one of them is located in front of the previous maximum, and the other behind it. Initially, the intensity associated with the central maximum increases, but later it subsides, and by the time corresponding to $j = 5$, the maximum disappears altogether. At the interval $5 < j < 8$, the laser pulse comprises two maxima where the intensity grows initially, but decreases subsequently. By the time corresponding to $j = 9$, the maxima in structure of the laser pulse are no longer present; the pulse is stretched and smoothed. At $9 < j < 11$, the intensities are low. Next, a central maximum reemerges followed by two peripheral maxima $(j = 12 - 15)$. The intensity in the central maximum increases and decreases later, and eventually (at $j = 16$), the central maximum merges with that located in front of it. After this, the pulse exhibits substantial asymmetry. The maximum that is closest to the laser pulse front is prevalent $(j = 17-18)$. The rear maximum continues to evolve: it breaks in two at $j = 19 - 20$.

The following conclusions stem from the simulations described above:

- The laser pulse remains symmetrical in η at the initial stage of propagation. The laser pulse asymmetry in η develops gradually beginning with j =10-12.
- Laser pulse disintegration occurs; the result is the emergence of a number of maxima in the intensity distribution. The maximal number of pikes in the example considered above is three;

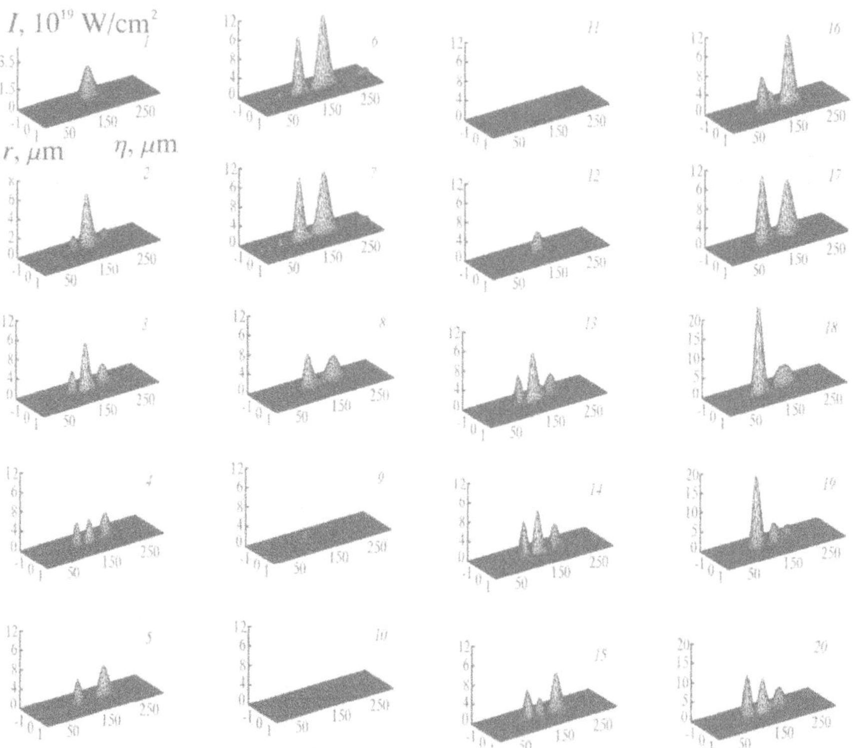

Fig. 9.6. The propagation of a laser pulse with an initially Gaussian temporal and radial profile in an unbounded plasma simulated using the wave equation with relativistic and charge-displacement nonlinearity. The maximal beam power is given by $P_0/P_{cr} = 1.4$ [115]

- The maxima in the intensity distribution exhibit pulsations, they emerge and disappear.
- Maxima can merge with each other.
- At some time intervals the maximal laser beam intensity decreases by an order of magnitude.

Therefore, simulations show that ultrashort laser pulses propagating into cold underdense plasmas undergo temporal self-modulation. Two characteristic spatial frequencies are associated with the solutions of the problem considered. The long wavelength scale δl_1 is due to the pulsations of a wave train comprising short pikes of midlevel intensity. These pikes are formed by the disintegration of a collapsed beam filament. In the case considered, $\delta l_1 \approx 200\,\mu\text{m}$. The short wavelength characteristic scale δl_2 is determined by the decay of the beam with the formation of a sequence of pikes with a certain characteristic interval between them. In the case discussed above, $\delta l_2 \approx 50$–$80\,\mu$m.

The time-integrated picture of the filament is determined by the following two circumstances. First, it is the pulsation of the entire beam having the characteristic scale of δl_1. Second, it is the small-scale temporal modulation of the intensity distribution inside the beam that is spread in space due to the finite lifetime of every particular pike. The pike lifetime is $\Delta t = 150$–$200\,$fs. During this time interval, the pike center passes a distance of $\Delta z = 45$–$60\,\mu$m. The pike size is typically $\delta z = 30$–$40\,\mu$m. Therefore, a particular pike must contribute a spot with the size of $\Delta z + \delta z = 75$–$100\,\mu$m to the time-integrated picture. Since the latter quantity is comparable to δl_2, the small-scale self-modulation should be distorted substantially in the overall filament structure.

9.3 Conclusions

A regime of intense, short laser pulse self-modulation in a cold underdense plasma is described. This regime is characterized by gradual suppression of the self-modulation; the final result is the dynamic formation of a two-dimensional soliton with extremely high power. Simulations demonstrate that giant broadening of the laser pulse spectra occurs at various distances of propagation in plasmas. The width of the laser radiation spectrum is inversely proportional to the duration of the pikes resulting from laser pulse self-modulation.

The following properties of solutions of the wave equation with relativistic and charge-displacement nonlinearity are found in computer simulations:

- Solutions exhibit substantial asymmetry in the co-moving variable η.
- Longitudinal self-modulation of propagating laser pulses occurs. Mathematically, this effect is attributed to the impact of the term with the second derivative in time, which is included in the model used.
- Laser pulse self-modulation begins when the first focus is formed. Note that models based on the nonlinear Schroedinger equation predict much less intense self-modulation and its beginning at a later time.
- After some propagation time, a prepulse emerges; a fraction of the laser radiation moves ahead of the entire pulse. There is an obvious tendency for laser pulse longitudinal stretching. Again, note that this effect is not described by nonlinear Schroedinger equation models.
- At near-threshold laser pulse powers, nonlinear self-focusing takes a fairly long time due to longitudinal stretching of the beam.

10. Propagation of Laser Radiation in Multiple-Stage Ionized Matter

In the previous chapters, we focused our attention on a number of aspects of laser–matter interactions at extremely high intensities. We considered preformed, completely ionized plasmas where laser radiation propagation is governed by relativistic and ponderomotive charge-displacement nonlinearities and the electromagnetic field self-effect due to plasma wave excitation. Yet, laser-irradiated plasma nonlinearity is not limited to these mechanisms. Additionally, the description of powerful laser pulse propagation in matter must include a theory of the ionization of matter by laser radiation and plasma optical polarization, i.e., the deformation of electron shells resulting in the emergence of oscillating dipole moments of multiply charged ions. Naturally, both of these phenomena contribute to the medium's dielectric response modification by propagating laser radiation.

The ionization of matter leads to the formation of a plasma cord where the "main" temporal part of the laser pulse propagates. The bell-shaped electron concentration profile causes laser radiation ionizational defocusing. Experimental studies and qualitative estimates for ionizational defocusing in gases were presented in [155–157]. In [158], the focusing of a short laser pulse with a maximal vacuum intensity of $10^{16}\,\mathrm{W/cm^2}$ into argon is treated in the case where the radiation is transported directly to the focal spot through the gas. Due to ionizational defocusing of the pulse transported, the maximal intensity decreases and the ionization stage does not get higher than two or three. In this situation, the laser pulse propagates in a slightly ionized gas so that the issues of propagation nonlinearity in multiple-stage ionized gases are not considered in [158].

A high ionization stage is possible in laser radiation vacuum transport to the focal spot. The intensities $I \approx 10^{18}\,\mathrm{W/cm^2}$ which are necessary for relativistic and charge-displacement self-channeling were obtained using this technique.

Numerical simulations of the nonlinear propagation of an intense ultrashort laser pulse in a gas under the conditions of multiple-stage ionization of its atoms are reported in [129]. Computations discussed in this work are performed for neon and lithium vapors. Also, issues of the development of optical-field-ionization X-ray lasers are analyzed in this paper. In this chapter, we follow the studies of [129]. Here we also consider the dipole moments of multiple-stage ionized heavy ions in intense fields using the Thomas–Fermi atom model.

10.1 General Description of Ionizational Defocusing

The propagation of a powerful laser pulse in an ionized gas is described by the nonlinear Schroedinger equation for field amplitude (envelope approximation) and the kinetic equation for relative ion concentrations in the emerging plasma [129]:

$$(v_{\mathrm{g}}^{-1}\partial_t + \partial_z)a + \frac{\mathrm{i}}{2k}\left[\Delta_\perp + k_{\mathrm{p}}^2\left(1 - \frac{\langle Z\rangle}{Z}\right)\right]a + \frac{1}{2}\mu^- a = 0\,, \tag{10.1}$$

$$\partial_t \boldsymbol{\alpha} = \mathbf{B}\boldsymbol{\alpha}\,, \qquad \boldsymbol{\alpha} = (\alpha_1, \alpha_2, \ldots, \alpha_{Z+1})^T\,, \tag{10.2}$$

$$\langle Z\rangle = \sum_{s=2}^{Z+1}(s-1)\alpha_s\,, \qquad \sum_{s=1}^{Z+1}\alpha_s = 1\,, \tag{10.3}$$

$$\mu^- = I^{-1}\partial_t(\rho\varepsilon_{\mathrm{ion}})\,, \qquad I = |a|^2\,, \tag{10.4}$$

$$\varepsilon_{\mathrm{ion}} = \frac{k_{\mathrm{B}}}{M}\sum_{s=2}^{Z+1}\sum_{n=1}^{s-1} J_n\alpha_s\,. \tag{10.5}$$

In (10.1), $\Delta_\perp = r^{-1}\partial_r + \partial_{rr}^2$ is the transverse Laplacian, $a(r,z,t)$ is the field complex amplitude, v_{g} is the group velocity, $k^2 = k_0^2 - k_{\mathrm{p}}^2$, where $k_0 = \omega/c$ is the vacuum wavenumber, $k_{\mathrm{p}} = \omega_{\mathrm{p}}/c$, ω_{p} is the plasma frequency (for fully ionized plasma), and Z is the nuclei charge number.

The average charge $\langle Z\rangle$ is given by (10.3); α_s are the relative concentrations of plasma ions in the s ionization stage. These concentrations satisfy kinetic equation (10.2). The ionization rate matrix is

$$\mathbf{B} = \begin{pmatrix} -\gamma_1 & 0 & 0 \ldots 0 & 0 & 0 \\ \gamma_1 & -\gamma_2 & 0 \ldots 0 & 0 & 0 \\ \ldots & \ldots & \ldots\ldots\ldots & \ldots & \ldots \\ 0 & 0 & 0 \ldots \gamma_{Z-1} & -\gamma_Z & 0 \\ 0 & 0 & 0 \ldots 0 & \gamma_Z & 0 \end{pmatrix}. \tag{10.6}$$

Both optical and collisional ionization contribute to the ionization rate $\gamma_s = \gamma_s^{\mathrm{r}} + \gamma_s^{\mathrm{e}}$, $s = 1,\ldots,Z$. Recombination can be neglected for ultrashort laser pulse propagation.

The ionization absorption rate μ^- is defined by (10.4). The ionization energy for a unit mass of gas is expressed by (10.5), where k_{B} is the Boltzmann constant, ρ is the gas density, M is the ion mass, J_n is the n-th stage ionization potential, and I is the radiation intensity.

Considering the laser pulse propagation in an unbounded medium for $t > 0$, we use the following natural initial and boundary conditions

$$a|_{t=0} = a_0(r,z)\,, \quad \boldsymbol{\alpha}|_{t=0} = \boldsymbol{\alpha}_0(r,z)\,, \tag{10.7}$$

$$a|_{r=\infty} = 0\,, \qquad a|_{z=\pm\infty} = 0\,, \qquad a_r|_{r=0} = 0\,. \tag{10.8}$$

The nonlinear Schroedinger equation (10.1) is applicable when the laser pulse amplitude varies slowly on the radiation wavelength scale, i.e., $c^{-1}|\partial_t a| \ll |ka|$.

Radiation diffraction, its refraction by the medium refractive index inhomogeneity due to the variation of the plasma free electron concentration resulting from the ionization, optical-field and collisional gas multiple-stage ionization, and ionizational absorption of radiation are included in the model represented by (10.1)–(10.8).

The next step is to specify the ionization model to be used with the equations presented above.

Threshold Ionization Model. There are several models of atom ionization by an exterior field. The particular feature of the problem considered is the need to describe ionization for a broad range of optical-field amplitudes and for ions at various ionization stages. In this respect, the threshold ionization model [153] is useful thanks to its relative simplicity, fairly universal character, and its adequacy for experimental data. The corresponding basic equations are

$$\gamma_s^{\mathrm{r}} = \gamma_0 \Theta\left(I - I_{\mathrm{thresh}}^{(s)}\right), \quad I_{\mathrm{thresh}}^{(s)} = 4\cdot 10^9 \frac{J_s^4}{s^2}\,, \quad s = 1,2,\ldots,Z\,, \tag{10.9}$$

where $\Theta(x)$ is the step function, J_s is the ionization potential, and γ_0 is a constant that does not depend on the ionization stage ($\gamma_0 \gg \tau_0^{-1}$, where τ_0 is the laser pulse duration). In the framework of this model it is assumed that instant ionization of an ion occurs when the radiation intensity is greater than the threshold value.

Tunneling Ionization Model. A quasi-classical formula for tunneling ionization of atoms and ions is proposed in [159]. The following version of this expression is used in the simulations discussed below:

$$\begin{aligned}
&\gamma_s^{\mathrm{r}} = \nu_0 \left(\frac{8.16}{\pi}\right)^{1/2} \frac{s^2}{n_s^{*4.5}} \left(\frac{10.88\, s^3}{\mathcal{E} n_s^{*4}}\right)^{2n_s^*-1.5} \exp\left(-\frac{2s^3}{3\mathcal{E} n_s^{*3}}\right), \\
&n_s^* = s\left(\frac{\mathrm{Ry}}{J_s}\right)^{1/2}, \qquad \mathcal{E}[\mathrm{AtomicUnits}] = 5.345\cdot 10^{-9}\left(I[\mathrm{W}\cdot\mathrm{cm}^{-2}]\right)^{1/2}, \\
&s = 1,2,\ldots,Z\,, \qquad \nu_0 = \frac{m_{\mathrm{e}} e_{\mathrm{e}}^4}{\hbar^3}\,.
\end{aligned} \tag{10.10}$$

In (10.10), e is the magnitude of the electron charge.

Collisional Ionization. A theory of the collisional ionization of neon atoms and ions is presented in [160]. Below, we use the following expression:

$$\gamma_s^{\mathrm{e}} = \begin{cases} N_{\mathrm{e}} c \sigma_s \,, & W > J_s \,, \\ 0 \,, & W \le J_s \,, \end{cases}$$

$$W = m_{\mathrm{e}} c^2 \left[\left(1 + \frac{I}{I_r} \right)^{1/2} - 1 \right], \qquad \sigma_s = \sigma_0 \left(\frac{\Omega_s}{2 l_{0s} + 1} \right) \left(\frac{\mathrm{Ry}}{J_s} \right)^2 \Phi(u_s) \,,$$

$$\sigma_0 = 0.8794 \cdot 10^{-16}\,\mathrm{cm}^2 \,, \quad \mathrm{Ry} = 13.606\,\mathrm{eV}, \quad u_s = \frac{W - J_s}{J_s} \,,$$

$$\Phi(u) = \frac{C_s u}{(u + \Phi_s)(u + 1)} \begin{cases} [u/(u+1)]^{1/2} \,, & s = 1 \,, \\ 1 \,, & s > 1 \,. \end{cases} \tag{10.11}$$

The relativistic factor is included in the above formulas. Here, I_r is the relativistic intensity. The values of constants are presented in Table 10.1.

Energy Conservation Law. Proceeding in a standard manner, we multiply (10.1) by a^* and add the result to its complex conjugate, thus arriving at

$$(v_{\mathrm{g}}^{-1} \partial_t + \partial_z)|a|^2 + \frac{\mathrm{i}}{2k}(a^* \Delta_\perp a - a \Delta_\perp a^*) + \mu^- |a|^2 = 0 \,. \tag{10.12}$$

Integrating the above equation from 0 to ∞ and using (10.5) and (10.7),

$$v_{\mathrm{g}}^{-1} \partial_t \int_{-\infty}^{\infty} \int_0^{\infty} |a|^2 r \mathrm{d}r \, \mathrm{d}z + \partial_t \int_{-\infty}^{\infty} \int_0^{\infty} \rho \varepsilon_{\mathrm{ion}} \, r \, \mathrm{d}r \, \mathrm{d}z = 0 \,. \tag{10.13}$$

Defining the laser pulse and ionization energies as

$$\varepsilon = 2\pi v_{\mathrm{g}}^{-1} \int_{-\infty}^{\infty} \int_0^{\infty} |a|^2 \, r \, \mathrm{d}r \, \mathrm{d}z \tag{10.14}$$

and

Table 10.1. Neon collisional ionization parameters

s	Shell	Ω_s	l_{0s}	C_s	ϕ_s
1	$1s^2 2s^2 2p^6$	6	1	16.9	1.46
2	$1s^2 2s^2 2p^5$	5	1	16.9	1.46
3	$1s^2 2s^2 2p^4$	4	1	16.9	1.46
4	$1s^2 2s^2 2p^3$	3	1	16.9	1.46
5	$1s^2 2s^2 2p^2$	2	1	16.9	1.46
6	$1s^2 2s^2 2p$	1	1	16.9	1.46
7	$1s^2 2s^2$	2	0	5.28	1.65
8	$1s^2 2s$	1	0	5.28	1.65
9	$1s^2$	2	0	7.2	2.56
10	$1s$	1	0	7.2	2.56

$$U = 2\pi \int_{-\infty}^{\infty} \int_{0}^{\infty} \rho \varepsilon_{\text{ion}} \, r \, \mathrm{d}r \, \mathrm{d}z \,, \tag{10.15}$$

respectively, we get the energy conservation law

$$\varepsilon + U = \text{const} \,. \tag{10.16}$$

Basic Equations in Co-Moving Variables. Similarly to the previous chapters, the following new variables are introduced to describe the propagation of spatially localized laser pulses: $\eta = v_{\mathrm{g}} t - z$, $\tau = t$. After this, the basic equations assume the form

$$v_{\mathrm{g}}^{-1} a_{\tau} + \frac{\mathrm{i}}{2k} \left[\Delta_{\perp} + k_{\mathrm{p}}^{2} \left(1 - \frac{\langle Z \rangle}{Z} \right) \right] a + \frac{1}{2} \mu^{-} a = 0 \,, \tag{10.17}$$

$$(v_{\mathrm{g}}^{-1} \partial_{\tau} + \partial_{\eta}) \boldsymbol{\alpha} = \mathbf{B} \boldsymbol{\alpha} \,, \tag{10.18}$$

$$\mu^{-} = I^{-1} (v_{\mathrm{g}}^{-1} \partial_{\tau} + \partial_{\eta}) (\rho \varepsilon_{\text{ion}}) \,. \tag{10.19}$$

The initial condition (10.7) for an initially focused Gaussian beam is written as

$$a|_{\tau=0} = a_0 \exp \left[-\frac{1}{2} \left(\frac{r}{r_0} \right)^2 - \frac{1}{2} \left(\frac{\eta - \eta_1}{\eta_0} \right)^2 - \mathrm{i} k \frac{r^2}{4 R_{\mathrm{f}}} \right] , \tag{10.20}$$

where R_{f} is the lens focal distance and r_0, η_0, η_1 are beam parameters. The boundary conditions retain the form given by (10.8).

10.2 Simulations of Ionizational Defocusing of Laser Pulses in Gases

Problem (10.17)–(10.20) was solved numerically using a spectral-finite-difference technique. General aspects of the numerical methods of this type are considered in [161, 162]. The model used describes the following phenomena:

- laser radiation diffraction;
- laser radiation refraction due to the medium's dielectric response variation related to the changes in free electron concentration $n_{\mathrm{e}}(r, z)$;
- generation of free electrons resulting from optical-field and collisional ionization;
- laser radiation collisional absorption owing to ionization.

Analytical estimates and simulations show that, depending on the ionizing gas density, the physics of laser radiation propagation in ionizing gases can be described adequately by one of several models discussed below.

1. In a low density gas, the medium's ionization has no significant impact on the character of the laser radiation propagation. The corresponding model applicability criterion is easy to derive for Gaussian beams. We must require that the optical path variation in a characteristic Gaussian caustic is much less than the laser radiation wavelength divided by 10: $\Delta nL \ll \lambda/10$. Since $\Delta n = \frac{1}{2}(\omega_p/\omega)^2$ and $L = \pi r_0^2/\lambda$, $n_e \ll 2m_e c^3/(e^2 r_0^2)$, or

$$n_e\,[\mathrm{cm}^{-3}] \ll 7 \cdot 10^{19} \left(\frac{10^{-4}}{r_0}\right)^2 . \tag{10.21}$$

Note that the above estimate is independent of the laser radiation wavelength.
2. For higher gas densities for which condition (10.21) is not met, the ionization effect on laser radiation propagation becomes significant and results in its additional refraction. In this case, the caustic length is shorter. The pertinent physics is different in gas and vacuum laser radiation transport to the focal spot.
3. Further gas density increase makes the role of laser radiation absorption important. Then, a depletion occurs at the laser pulse front.

Below, the above situations are illustrated by specific examples.

Low-Density Gas Ionization by Short Laser Pulses. Consider the focusing of an ultrashort laser pulse with wavelength $\lambda = 0.4\,\mu\mathrm{m}$, focal spot aperture $r_0 = 10\,\mu\mathrm{m}$, maximal intensity $I = 2\times10^{17}\,\mathrm{W/cm^2}$, and longitudinal size $\eta_0 = 15\,\mu\mathrm{m}$ in the vapors of lithium having density $N_0 = 5 \times 10^{18}\,\mathrm{cm}^{-3}$. The corresponding solution is depicted in Fig. 10.1. In particular, the two-dimensional spatial charge distribution which remains behind the propagating laser pulse is shown in this figure. In Fig. 10.1(a) the distance between the laser pulse and the focal spot is -0.31 cm; in Fig. 10.1(b) the laser pulse traverses the focal spot, and in Fig. 10.1(c) the entire laser pulse has gone through the Gaussian caustic. In this simulation, the medium's density is low and the emergence of free electrons does not influence the character of the laser pulse propagation.

Ionizational Defocusing of a Laser Pulse Transported to the Lens Focal Plate via a Gas. This case was studied in [158]. The equations treated in that paper are similar to those of this chapter. The propagation of focused laser pulses with intensities up to $I = 2 \times 10^{17}\,\mathrm{W/cm^2}$ in a semi-infinite argon gaseous medium was simulated.

Simulations demonstrate that the laser pulse does not reach the theoretically predicted focal plate but undergoes a spreading prior to reaching it due to ionizational defocusing. A diffraction ring is formed as the laser pulse is diffused. The focal caustic length decreases by a factor of several, and the ionization stage is not greater than 2.

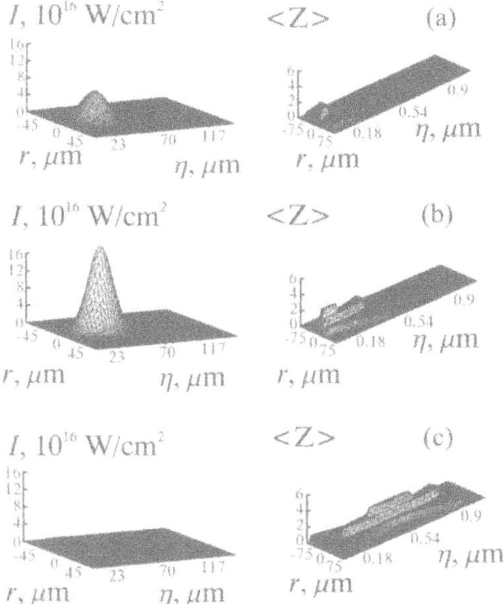

Fig. 10.1. Simulation of ultrashort laser pulse propagation in lithium vapor: laser radiation intensity and ionized plasma average charge distributions. **(a)** The distance between the laser pulse and the focal spot is -0.31 cm; **(b)** the laser pulse goes through the focus; **(c)** the laser pulse has gone through the Gaussian caustic [129]

Small Scale Ionizational Phase Defocusing of Laser Radiation in a Multiple-Charge Plasma for Vacuum Energy Transport to the Focal Spot. Vacuum laser energy transport to the focal spot is a method of increasing the radiation intensity in it [163]. In this case, the character of ionizational defocusing was investigated in [129]. Consider the propagation of an Nd-glass laser pulse with maximal intensity $I = 2 \times 10^{17}\,\mathrm{W/cm^2}$, and the transverse and longitudinal sizes are $r_0 = 10\,\mu\mathrm{m}$ and $\eta_0 = 12.5\,\mu\mathrm{m}$ respectively in neon; the density of the latter is $N_0 = 3 \times 10^{19}\,\mathrm{cm^{-3}}$. The evolution of the laser pulse spatially two-dimensional intensity distribution (in the co-moving variables) is shown in Fig. 10.2 for laser pulse displacement from its initial location $\delta z = 5.2 + 10.5 \times \mathrm{j}\,\mu\mathrm{m}$, where $\mathrm{j} = 1, \ldots, 6$. The average charge distribution along the laser pulse propagation axis at times given by τ=4, 35, 44, and 61 fs is presented in Fig. 10.3 (also in co-moving variables).

A substantial feature of the propagation process considered is small-scale ionizational phase defocusing. The small-scale character of the phenomenon is determined by the fact that multiple-stage ionization occurs in the gas. The electron concentration profile resulting from gas ionization is a columnar chart-shaped function of η and τ since both optical field and collisional ionization have threshold character. This fact is illustrated by Figs. 10.1 and 10.3. In a way, one can say that the laser radiation goes through a step-shaped phase grating. As a result, a complicated interference pattern emerges including a

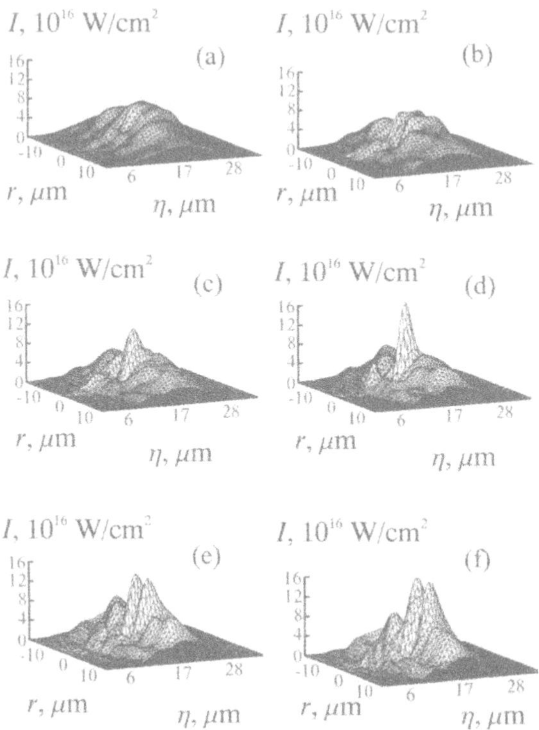

Fig. 10.2. The ultrashort laser pulse propagation in atmospheric pressure Neon. $N_0 = 3\times10^{19}\,\mathrm{W/cm^2}$. The threshold ionization model is used. Laser radiation intensity distributions corresponding to the pulse propagation distances of **(a)** 15.7 μm, **(b)** 26.2 μm, **(c)** 36.7 μm, **(d)** 47.2 μm, **(e)** 57.7 μm, and **(f)** 68.2 μm are shown [129]

large number of maxima and minima. The laser pulse is efficiently scattered to the periphery. Note that the numerical solution of a set of equations of the type represented by (10.17)–(10.20) is complicated by the discontinuity of its solutions, and the treatment procedure involves smoothing the equation coefficients and using a nonuniform grid with a higher number of points in the areas where efficient ionization occurs.

It can be shown that the Rayleigh range L_0 in a medium where Z-stage ionization has taken place is reduced by a factor of Z^2. The laser pulse aperture is divided into Z transverse zones; their average size is r_0/Z. For a particular zone, the Rayleigh range is

$$L \simeq \frac{\pi(r_0/Z)^2}{\lambda} = \frac{1}{Z^2}\left(\frac{\pi r_0^2}{\lambda}\right) = \frac{1}{Z^2} L_0 \,. \tag{10.22}$$

Consequently if criterion (10.21) is met, the laser pulse is focused into a spot with transverse size r_0 and the caustic length L_0 causes a Z-stage ionization and gets scattered by perturbations with the transverse size r_0/Z, and the

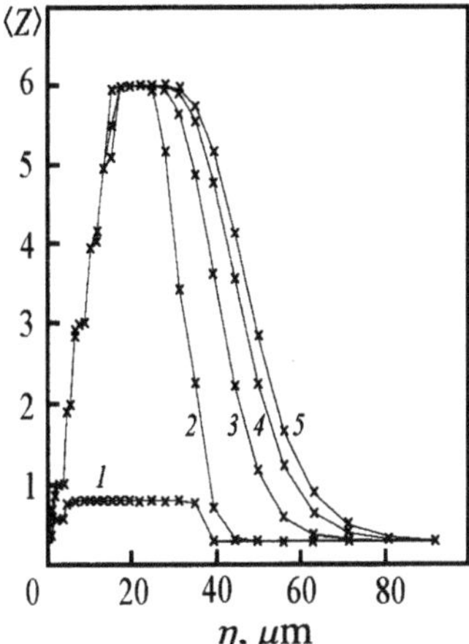

Fig. 10.3. Average charge distribution at the axis of ultrashort laser pulse propagation in atmospheric pressure neon in co-moving variables at consecutive times: **(1)** 4 fs; **(2)** 35 fs; **(3)** 44 fs; **(4)** 61 fs; **(5)** 72 fs. The threshold ionization model is used

propagation distance is Z^2 times shorter. The scattering angle is Z times greater: $\theta \simeq \lambda Z/r_0$. Note that small-scale ionizational phase defocusing can be an obstacle to laser pulse self-focusing in matter.

Simulations using both threshold and tunneling ionization models are reported in [129]. The former performs poorly far from the ionization threshold, whereas the latter is applicable far from the ionization threshold but reflects the near-threshold cross section behavior inadequately. Besides, the precision of the tunneling ionization model tends to be low for intensities higher than $I = 2 \times 10^{15}\,\mathrm{W/cm^2}$. Small-scale ionizational defocusing is predicted in the framework of both models. Cross section smoothing near the ionization threshold is characteristic of the tunneling model which leads to the prediction of smoother $n_e(r, z)$ distributions. As a result, the second model yields small-scale phase distortions at distances twice as great as those predicted by the first.

Collisional ionization can be neglected for the densities employed in the above simulations. Its role becomes more important for higher gas densities. Note that here we also neglected collisional ionization from the excited states of atoms and ions.

Under certain conditions, laser radiation absorption due to ionization is significant. This absorption causes depletion at the laser pulse front. This effect can be seen in Fig. 10.2. For higher intensities, absorption is not as important since the laser pulse scattering is a faster process than absorption.

The basic conclusion which can be drawn from the above studies is as follows: to implement the optical field ionization X-ray laser scheme using a short powerful generating laser, condition (10.21) should be provided for to avoid any significant effect of the medium on the propagating laser radiation.

When criterion (10.21) is not met, which might be the case for laser pulse vacuum transport to the focal spot, small-scale ionizational phase defocusing occurs due to the columnar chart-shaped electron concentration distribution. The situation where the scattering angle increases by a factor of Z and the Rayleigh range becomes Z^2 times shorter is unacceptable for OFI X-ray laser scheme implementation. Laser pulse front depletion can be significant under some circumstances. The predicted laser pulse scattering distances differ by approximately a factor of 2 when the threshold and tunneling ionization models are used, thus making it important to improve the ionization models.

Finally, we must remark that the inequality $P_0 < P_{\mathrm{cr}}$, where P_{cr} is the relativistic and charge-displacement self-channeling critical power, held true for the conditions simulated. Another relevant issue is the potential impact of Kerr nonlinearity stemming from the polarization of ion shells by the intense optical field on the character of the laser radiation propagation.

10.3 Experimental Demonstration of Ionization-Induced Defocusing of Short-Pulse, High-Power Lasers in Gases

The defocusing of a short-pulse high-power laser was first observed by Monot and co-workers [155]. It was shown that a 1-ps, 1-TW laser pulse focused in a gas cell is strongly defocused by the radial electronic density gradient of the plasma created on the rising edge of the pulse well before reaching the focus position in vacuum, limiting the laser intensity which can be obtained in the medium. A study of the laser energy transmitted through argon gas as a function of the backing pressure (that is, the pressure before the jet) has shown that the transmission decreases dramatically as the pressure is increased. It drops from 100% at 10 mbar to 15% at 1 bar for 0.6-TW incident laser power (see Fig. 10.4).

This behavior was interpreted as an increase of beam divergence with the pressure and then of the energy refracted out of the cone angle of the collecting optics used to collimate the output beam. At the same time, images of the plasma fluorescence in the visible at 90° from the direction of propagation of the laser showed that the focus moves backward with respect to the direction of propagation of the laser when the gas pressure is increased. This result indicates that ionization-induced defocusing happens further and further away from the vacuum focus position as the gas pressure is increased and therefore, reduces more and more the maximum intensity that can be achieved in the medium. Investigation of the dependence of transmission with the atomic number of the gas target [165] shows that the transmitted energy is higher for lighter elements at a given pressure. For a backing pressure of

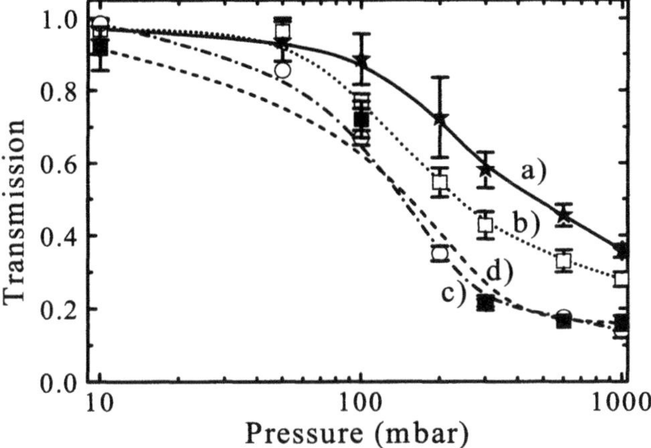

Fig. 10.4. Transmission of a 1-ps, 1.053-μm laser pulse as a function of the backing pressure of argon, for P= 3×10^{-3} TW (**a**), 10^{-2} TW (**b**), 10^{-1} TW (**c**), and 6×10^{-1} TW (**d**) [155]

1 bar, the transmission of a 1-ps, 0.1-TW pulse is still 30% in helium gas, whereas it is only 15% in argon. This behavior is explained by the smaller number of electrons released by light elements than by heavy materials and by the threshold ionization intensity which decreases with atomic number. Under a few assumptions, the authors deduced from these measurements a maximum intensity that can be achieved in helium and argon gases as a function of the backing pressure. It was demonstrated that the intensity tends asymptotically to the ionization intensity of the first charge state of the gas target, as expected. Figure 10.5 depicts the maximum intensity which can be reached as a function of the backing pressure for argon and helium.

These observations, it was found, are in good agreement with the numerical simulations of Rae [166] and with the analytical results of Fill [167]. Further investigations of ionization-induced defocusing are reported in [168–171]. Particularly, Nikitin and co-workers [170] showed that ionization-induced defocusing was responsible for the erosion of the rising edge of the laser pulse, which leads to pulse shortening. Finally, recently, Chessa and co-workers [171] studied the angular distribution and the temporal evolution of a pulse undergoing ionization-induced refraction, using the time-frequency correspondence of a 120-fs Ti:sapphire laser, linearly chirped to 1.8 ps. The experiments were carried out in 375 mbar of helium with a maximum laser intensity in vacuum of 4×10^{16} W/cm^2. It was shown that the front part of the output pulse, whose energy remains below the ionization threshold intensity of neutral helium, keeps a maximum intensity on-axis, whereas the part of the pulse which energy is above the ionization intensity of the gas is defocused by the radial electron density gradients and exhibits a ring shape.

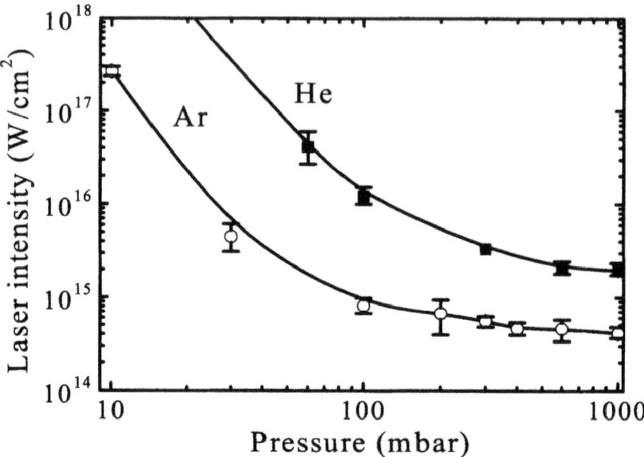

Fig. 10.5. Maximum laser intensity reached in a gas as a function of the backing pressure for argon and helium [165]

10.4 Spectral Blueshifting of Short-Pulse, High-Power Lasers in Gases

The spectral blueshifting of a short-pulse, high-intensity laser propagating in a gas is caused by rapid temporal variation of the index of refraction of the medium undergoing ionization. The wavelength shift for an underdense plasma ($n_e \ll n_c$) is given by

$$\delta\lambda = \frac{-n_0 e^2 \lambda^3}{8\pi^2 \epsilon_0 m c^3} \int_0^z \frac{\mathrm{d}Z(s)}{\mathrm{d}t} \mathrm{d}s \,, \tag{10.23}$$

where the integration is performed over the interaction length. It is a self-phase modulation effect, as in optical fibers, with an explicit time dependence of nonlinearity. The asymmetry of the spectrum after interaction with the gas target is due to the irreversibility of the ionization process with respect to time, i.e., the medium is ionized on the leading edge of the pulse and does not recombine in the pulse duration. In other words, $\mathrm{d}Ne/\mathrm{d}t > 0$ in the pulse duration. First suggested by Bloembergen, spectral blueshifting of a laser propagating in a rapidly ionizing gas was observed in the early 1970s by Yablonovitch with a nanosecond CO_2 laser [172]. In this case, the plasma was created by collisional ionization of the gas target. The first observations of spectral blueshifting and broadening with femtosecond pulses were reported by Wood and co-workers in the 1980s [173, 174], followed by numbers of publications on the topics (see, for example, [175–178]). The physical process leading to the creation of a plasma was then optical-field ionization (OFI). In [174], the results of a pump/probe experiment with a probe copropagating with the pump are presented. The spectrum of the probe, recorded for different pump/probe delays, provides time-resolved information on the

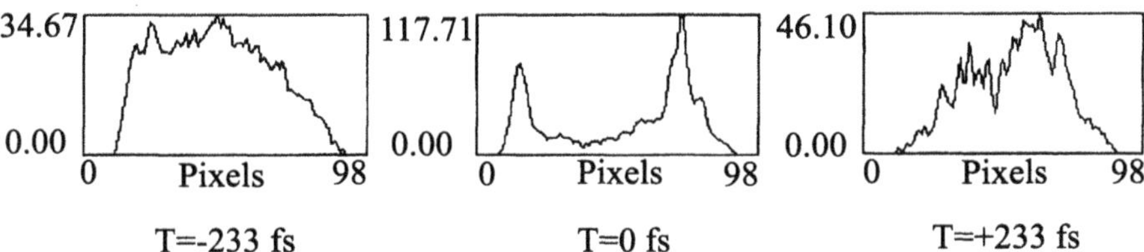

Fig. 10.6. Spatial evolution of the output spectrum of the probe for a −233, 0, and +233 fs delay [178]

blueshift with a resolution given by the probe pulse duration. In this way, it is shown that the blue wing of the pulse occurs in the very early part of the pulse, once the laser intensity exceeds the ionization intensity of the gas target, and ionization-induced defocusing begins. Space- and time-resolved two-color pump/probe experiments, described in [178], show that ionization-induced defocusing makes spectral blueshifting spatially dependent. In these experiments, a KrF ($\lambda = 248\,\mathrm{nm}$), 400-fs laser, used as a pump, was focused at intensities in vacuum between 10^{14} and $10^{15}\,\mathrm{W/cm^2}$ in a cell filled with argon or air at pressures up to 25 bars. A second pulse with a 100-fs duration and a 497-nm wavelength was copropagated with the pump. The spatially-resolved spectrum of the probe pulse after interaction with the plasma created by the pump was recorded as a function of the pump–probe delay. Figure 10.6 depicts the spatial evolution of the output spectrum of the probe for delays between -233 and +233 fs, when the pump and the probe were focused in air.

One can see that at negative delay times, the spectrum of the probe is unshifted since it precedes the pump and therefore interacts with the gas. When the pump and the probe overlap in time, the bleushifted component has a ring shape due to temporal and spatial variations of the index of refraction in both radial and axial directions. For positive time delays, the spectrum of the probe no longer has a shifted component but is spatially inhomogeneous due to propagation and refraction in the plasma. On the other hand, experiments performed in the same laboratory with the pump alone and presented in [175], show that the spectrum exhibits modulations that are attributed by the authors to interferences between frequency components occurring at different times in the laser pulse.

10.5 Thomas–Fermi Atom in an Intense Field

As discussed in the previous section, intense laser radiation propagating in a gas causes multiple-stage ionization of its atoms. The strong electromagnetic field induces deformations of the electron shells of ions, and oscillating dipole moments of atoms emerge; the result is the medium's optical polarization and a corresponding variation of its dielectric response.

Calculating of the nonlinear dipole moments of multiple-charge ions induced by an intense field is a complicated problem. The quantum mechanical methods for intensities from the 10^{17}–10^{20}W/cm^2 range are just being developed, and the atomic data necessary for calculating the optical polarization of the multiple charge ions plasma are not available. However, there is a relatively simple and efficient technique for estimating the multiple-charge ion dipole moments, their ionization stages, and several other parameters in intense fields, namely the Thomas–Fermi atom model. This method applies particularly well to heavy atoms with a large number of bounded electrons. The one-dimensional Thomas–Fermi atom model is described, for example, in [424]. This model becomes two-dimensional in an exterior field. Below, we present this model (atomic units are used in the derivations to follow).

Consider an atom comprising a point nucleus with a positive charge Z and a negatively charged electron cloud surrounding it. The Thomas–Fermi theory is based on the following assumptions. First, it is assumed that the number of electrons is large enough to justify a continuous description of the electron cloud. Second, the atomic electrons are treated as a degenerate electron gas with zero temperature. Third, the scalar potential distribution is described by the Poisson equation. Fourth, it is assumed that the bounded system total energy must be minimal. A consistent theory of an atom can be developed on the basis of the above assumptions.

Let us examine the basic equations of the model discussed. In the space around the atomic nucleus, the scalar potential $\varphi(\boldsymbol{x})$ obeys the Poisson equation,

$$\Delta\varphi = -4\pi\rho = -4\pi Z\delta(\boldsymbol{x}) + 4\pi\rho_e(\boldsymbol{x}), \tag{10.24}$$

and the electron cloud negative charge is

$$Q_e = \int \rho_e \, \mathrm{d}^3x . \tag{10.25}$$

As mentioned above, the electron density ρ_e and the charge Q_e are viewed as continuous in the framework of the model used.

Now consider the system's total energy. Its minimization makes it possible to establish the relation between φ and ρ_e which is necessary for turning (10.24) into a closed model. The nucleus potential is

$$\varphi_z = \frac{Z}{|\boldsymbol{x}|} . \tag{10.26}$$

The potential due to the electron cloud is

$$\varphi_e = -\int \frac{\rho_e(|\boldsymbol{x}'|)}{|\boldsymbol{x}-\boldsymbol{x}'|} \, \mathrm{d}^3x' . \tag{10.27}$$

For the electron cloud–nucleus interaction energy,

$$W_{Ze} = -\int \rho_e\varphi_Z \, \mathrm{d}^3x , \tag{10.28}$$

and the energy of the interactions between electrons is

$$W_{ee} = -\frac{1}{2}\int \rho_e \varphi_e \, \mathrm{d}^3 x \,. \tag{10.29}$$

The kinetic energy of electrons at absolute zero is expressed as

$$T_e = \chi \int \rho_e^{5/3} \, \mathrm{d}^3 x \,, \qquad \chi = \frac{3}{10}(3\pi^2)^{2/3} \,. \tag{10.30}$$

So, total atomic energy is

$$E_\Sigma = W_{Ze} + W_{ee} + T_e \,. \tag{10.31}$$

Consider the conditional extremum problem. The relation between φ and ρ_e minimizing the total energy with the condition $Q_e = \text{const}$ must be found. The corresponding variational equation is as follows

$$\delta(E_\Sigma + \varepsilon Q_e) = 0 \,, \tag{10.32}$$

where ε is a Lagrange multiplier. Calculating the variation using (10.22)–(10.30), one finds easily that

$$\rho_e = \frac{2^{2/3}}{3\pi^2}(\varphi - \varepsilon)^{3/2} \,, \qquad \varphi = \varphi_Z + \varphi_e \,. \tag{10.33}$$

When there is an exterior field, (10.33) holds true as well, but $\varphi = \varphi_Z + \varphi_e + Fx_3$. Finally, the model of the two-dimensional Thomas–Fermi atom in an exterior field is (10.24)

$$\rho_e = \frac{2^{2/3}}{3\pi^2}(\varphi + Fx_3 - \varepsilon)^{3/2} \,. \tag{10.34}$$

The Lagrange multiplier ε is interpreted as the atom ionization potential. It is chosen to provide for satisfying condition (10.25) by solving (10.34). Integration is performed over the atom volume which is defined by the condition

$$\varphi + Fx_3 - \varepsilon > 0 \,. \tag{10.35}$$

The problem can be made independent of the nucleus charge Z by the following change of variables: $\tilde{\boldsymbol{x}} = \boldsymbol{x}/x$, $\tilde{\varphi} = \varphi/\varphi_0$, $\tilde{F} = F/F_0$ where $x_0 = Z^{-1/3}$, $\varphi_0 = Z^{4/3}$, $F_0 = Z^{5/2}$.

Obviously, condition (10.25) defines a function

$$Q_e = Q_e(\varepsilon, F) \,. \tag{10.36}$$

Studies of the potential for the two-dimensional problem show that for a given field F, a maximal value Q_e^m corresponds to a minimal value of the ionization potential ε^m. After a further decrease in ε, the atom stops existing as a localized system, and the entire problem no longer makes sense. The quantities $Z - Q_e^m(F)$ and $\varepsilon^m(F)$ are the atom ionization stage and potential corresponding to the field F.

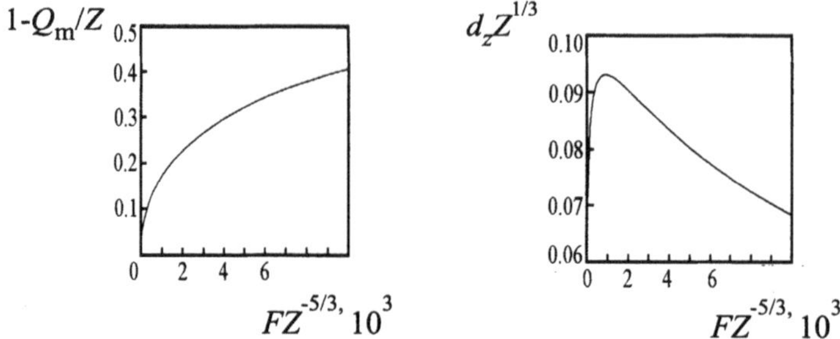

Fig. 10.7. The dependence of the Thomas–Fermi atom ionization stage and dipole moment on the exterior field

The dipole moment of an atom in a field F can be calculated with the help of the present model. It is directed along the $\boldsymbol{e}_3$ axis, and its magnitude is

$$d_3(F) = \int \rho_e(\boldsymbol{x}) x_3 \, \mathrm{d}^3 x \,. \tag{10.37}$$

The dependencies of the atom ionization stage and dipole moment on the exterior field are depicted in Fig. 10.7. The bell-shaped form of the curve for the dipole moment is explained as follows: at low intensities, the field effect on the atom is also low, and at high intensities, the atom is ionized and its volume decreases, which results in reducing the dipole moment as well.

The calculation presented above is performed for a time-independent field. For a time-dependent field of the form $F(t) = F^0 \cos(\omega t)$, the same procedure should be carried out using the maximal amplitude value F^0. Then, the ionization stage should be assumed constant, and the solution procedure should be repeated with a lower field amplitude. Generally, this allows finding the dipole moment dependence on time. However, such a calculation is hardly feasible due to the approximate character of the model.

11. Experiments on Laser–Matter Interaction in the Relativistic Regime

By using the chirp pulse amplification (CPA) technique [4, 5], considerable progress has been made developing compact and short pulse powerful lasers in the past 10 years (see Fig. 1.1). The continuous increase in the performance of these lasers makes it possible now to study the relativistic regime of interaction which is somewhat artificially defined by $a = 0.85 \times 10^{-9} \times (\lambda[\mu\text{m}]) \times (I_{\max}[\text{W/cm}^2])^{1/2} \geq 1$, where a is the normalized vector potential. Intensities as high as $10^{20}\,\text{W/cm}^2$ were recently reached [179]. Such intensities correspond to $a = 8.5$ and to electric fields about four decades higher than the electric field atomic unit.

Until now, two materials were mainly used as lasing media to generate high-power ($P > 1\,\text{TW}$) short pulses. The first is the Nd:glass material which recently allowed reaching 1.25 PW on the NOVA laser system of the Lawrence Livermore National Laboratory [180]. It is pumped by flash lamps; this limits the repetition rate to a few shots per hour. On the other hand, the spectral bandwidth limits the pulse duration to a few hundred femtoseconds ($\sim 300\,\text{fs}$). Other systems with lower power are currently in use in the United States at the Naval Research Laboratory [181], at the University of Rochester [182], and at the University of Michigan [183]; in Japan at the University of Osaka (GEKKO-XII) [184]; in France at the CEA at Limeil–Valenton (P102) [185] and at the Ecole Polytechnique at Palaiseau [186]; in the United Kingdom at the Rutherford Appleton Laboratory (VULCAN) [187]; and in Germany at the Max Born Institute in Berlin [188]. Figure 11.1 shows the main existing Nd:glass systems as well as some projects of petawatt lasers.

The other material commonly used to produce high-power short pulses is the Ti:sapphire. Due to its large spectral bandwidth, this lasing medium allows generating extremely short pulses. Pulses as short as 5 fs have been recently achieved [189]. Another interest in using Ti:Sapphire is that it is pumped by lasers; this permits repetition rates as high as 1 kHz [190, 191]. Therefore, the short pulse duration and high repetition rates allow one to develop tabletop size multi-TW lasers; a large number of these systems are routinely used worldwide. The highest power reached until now with a Ti:sapphire laser is 200 TW. Extrapolation to higher power, e.g., to the PW level, seems to be difficult because of the prohibitive size (and cost) of the

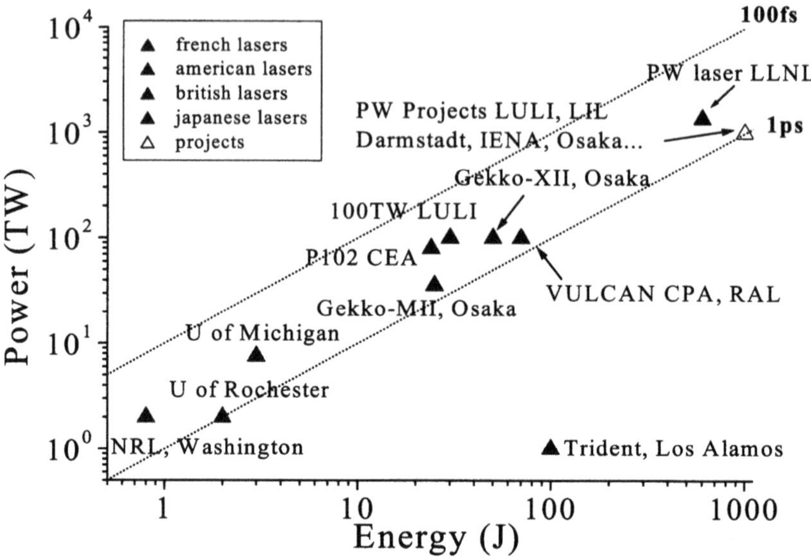

Fig. 11.1. TW Nd:glass laser systems around the world

pumping lasers. Figure 11.2 shows a synthesis of the main Ti-sapphire laser systems with power in the 1–100 TW range.

11.1 Enhancement of Self-Channeling Distance by an Exterior Supply of Energy

A number of applications that we shall describe in the next sections, like optical-field-ionized X-ray laser schemes (see Sect. 11.2) or laser wakefield accelerators (see Sect. 11.4) require a powerful short-pulse laser to interact with a neutral gas or a high-density plasma ($n_e \geq 0.01 \times n_c$) and high-intensities ($I > 10^{15}\,\mathrm{W/cm^2}$) over distances ranging from a few millimeters to several centimeters. However, as discussed in Chap. 10, a high-power, short-pulse laser propagating in a high-pressure gas undergoes strong defocusing due to the plasma created on the leading front of the pulse, once the intensity exceeds the threshold ionization intensity. On the other hand, it was shown in the previous chapters of this book that in the relativistic regime of interaction, numerous theoretical and experimental works have demonstrated that both relativistic and ponderomotive self-channeling can significantly enhance the interaction length. However, the interaction length can still be increased by using a plasma created by an exterior supply (a laser [192–198] or a capillary discharge [199–201]) with a tailored radial electron density profile (with a density minimum on-axis) or microcapillaries [202–205].

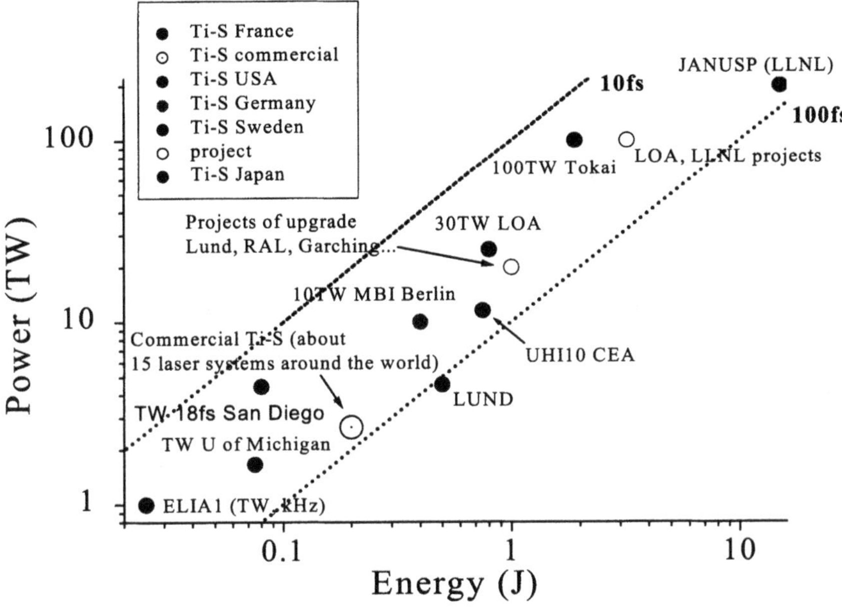

Fig. 11.2. TW Ti:sapphire laser systems around the world

One of the most commonly used exterior supplies to create plasma waveguides is a relatively long-pulse laser (0.1–1 ns) with moderate energy (0.1–1 J) focused at intensities between 10^{13} and 10^{14} W/cm^2. Durfee and Milchberg [192] were the first to demonstrate optical guiding of laser pulses with moderate power (up to 0.4 GW) over a distance of more than 20 Rayleigh lengths in a laser-produced plasma channel, using a two counterpropagating beam configuration. The channel was created by shock-driven radial expansion of a plasma several millimeters long, produced by focusing a first laser beam with an axicon (an axiconic lens) [206] into a gas chamber filled with up to 30 Torr of xenon. The second pulse was injected after an appropriate delay into the plasma in the direction opposite to the first pulse.

Using the same experimental arrangement, Nikitin and co-workers [195] demonstrated guiding 90-fs pulses from a Ti:sapphire laser system at intensities up to 5×10^{15} W/cm^2 over a distance of 1 cm. The coupling efficiency, i.e., the measured fraction of incident laser energy transmitted through the plasma channel, was 30%. When the plasma waveguide was produced in an ambient background pressure, the guided intensity was found limited by the strong refraction of the injected pulses due to optical-field ionization of the neutral gas far from the channel entrance. In recent experiments, Nikitin and co-workers [196] solved this problem by using a pulsed gas jet configuration [197]. They demonstrated guiding pulses of maximal power up to 0.3 TW in 1.5-cm long preformed plasma waveguides. Coupling of up to 50% was measured, and guided intensities up to 5×10^{16} W/cm^2 were achieved.

In this experiment, the guided intensity was limited by the presence of long-scale-length (500 μm) longitudinal density gradients on the edges of the jet. As the electron density dropped on both ends of the waveguide, the efficiency of inverse bremsstrahlung heating and ionization was reduced. Therefore, the plasma expanded more slowly in this region than in the middle of the jet, and the delay before injection had to be increased to achieve efficient coupling. Consequently, rather large spot diameters were obtained, reducing the maximum guided intensity.

Another technique for creating plasma waveguides was recently proposed by Volfbeyn and co-workers [198]. This so-called "igniter-heater" technique uses a two-beam configuration. The igniter, a short-pulse (75-fs), low-energy (20-mJ) beam is brought to line focus using a cylindrical retroreflector to ionize a gas jet. The heater pulse (160 ps, 280 mJ) is subsequently used to heat the existing plasma by inverse bremsstrahlung. The plasma waveguide is then created by shock-driven expansion. The use of an "igniter" pulse reduces the energy requirements for the "heater" pulse and allows better control of plasma parameters and reproducibility shot-to-shot than a single-beam configuration. Slab and cylindrically symmetrical channels were produced, depending whether the two pulses were copropagated or were propagated orthogonally to each other. Preliminary experiments on guiding of a 0.5-TW (75-fs, 40-mJ) beam focused at $5 \times 10^{17}\,\mathrm{W/cm^2}$ in a 1-mm slab channel demonstrated only 20% transmission ($I_{\mathrm{out}} \sim 10^{16}\,\mathrm{W/cm^2}$) because of important leakage losses due to the particular geometry of the plasma channel; the guiding occurred in only one dimension.

Slow capillary discharges were also used to create plasma waveguides for high-power, short laser pulses [199–201]. A plasma is produced by the evaporation of the capillary walls due to an electrical discharge. In a slotted capillary, a plasma jet is obtained which flows in vacuum, and the laser is focused onto the jet. The plasma temperature and density can be controlled by varying the discharge parameters, and tailored density profiles for laser pulse guiding can be achieved by changing the capillary geometry and/or material. Guiding of a 0.3-TW (100-fs, 30-mJ) Ti:sapphire laser on a distance of 10 mm was demonstrated by this method. The guided intensity was estimated at $10^{16}\,\mathrm{W/cm^2}$. Experiments with an axisymmetric capillary were also carried out by Ehrlich and co-workers [200,201]. In that case, the laser was injected in a 600-μm internal diameter capillary and was propagated through the plasma generated from material ablated onto the inner wall of the capillary by the discharge. Guided intensities greater than $10^{16}\,\mathrm{W/cm^2}$ were achieved over distances up to 6.6 cm ($\sim 75\times$Rayleigh length). Figure 11.3 shows the beam pattern without (left views) and with (right views) guiding through 3-cm (a) and 6.6-cm (b) long capillaries. Laser defocusing due to optical-field ionization of the weakly ionized capillary plasma was identified as a limiting factor in achieving higher intensities and/or guiding very high-power ($P \gg 0.3\,\mathrm{TW}$) laser pulses.

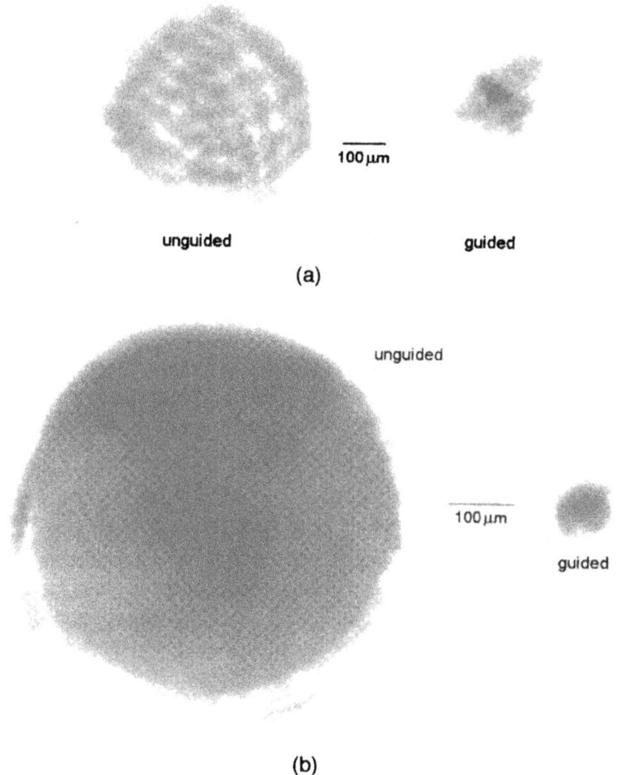

Fig. 11.3. Beam pattern without (*left views*) and with (*right views*) guiding through 3 cm (**a**) and 6.6 cm (**b**) long capillaries [201]

Efficient guiding of high-power laser pulses was also reported by Jackel and co-workers [202], later on, by Borghesi and co-workers [203], and by Dorchies and co-workers [204], using hollow glass microcapillary tubes. Experiments by Jackel and co-workers, performed with a 1-TW laser focused at $10^{17}\,\mathrm{W/cm^2}$ at the entrance of a 100–200 µm microcapillary tube, show that for energies below the breakdown threshold, laser guiding is due to multiple grazing-incidence Fresnel reflections on the inner surface of the microcapillary tube. For intensities above the breakdown threshold, the laser beam is reflected on the dense plasma created at the glass tube wall by the pedestral or the rising edge of the pulse. In this case, propagation of the pulse occurs through reflections on the plasma at density $n_\mathrm{e} = n_\mathrm{c} \times \cos^2\theta$, where n_c is the critical density and θ is the angle of incidence. The transmission through the microcapillary tube follows the simple law $T = T_\mathrm{ins}(R_\mathrm{bounce})^N$, where T_ins is the coupling efficiency, R_bounce is the plasma reflectivity per bounce, and N is the number of bounces undergone by the laser pulse over the entire length of the tube. A 1-TW laser was guided in a 3-cm long, 100-µm diameter microcapillary tube. The maximum intensities at the input and the output of the tube were 10^{17} W/cm^2 and 10^{16} W/cm^2, respectively. The transmission, it was found, decreases with laser energy and with the diameter and the length of the tube. A mode structure, it was observed, develops in the waveguide with a number of modes, depending on the diameter of the tube.

The experiments by Borghesi co-workers were carried out with the VULCAN laser system [187] at the 10-TW power level. Guiding through microcapillary tubes with 40 and 100-μm diameters and lengths up to 1 cm was observed; transmission was greater than 80%.

In experiments by Dorchies and co-workers, a 30-mJ, 120-fs laser pulse was guided within a single mode over 10 cm (100×Rayleigh length) through a hollow microcapillary tube whose inner diameter was between 45 and 70 μm. No breakdown of the internal wall of the tube was observed, and guided intensities up to 10^{16} W/cm^2 were demonstrated. Coupling efficiency as high as 97% and transmission as high as 50% were obtained for the tube with the largest diameter (70 μm). Experiments with gas-filled microcapillary tubes were also carried out at helium pressures up to 40 mbar. In that case, pulse shortening as well as a transmission decrease and spectral broadening were observed as a consequence of the optical-field ionization of the gas by the laser pulse.

11.2 X-Ray Laser

In the current schemes of X-ray lasers, the desired ionization state is reached through collisions with free electrons heated by a large optical driving laser. The size and the cost of these driving lasers put them out of the range of most laboratories. With the development of tabletop, high-intensity, short-pulse lasers, new X-ray laser schemes based on rapid three-body recombination following optical-field ionization (OFI) of a gas target are the object of numerous theoretical works [206–211]. Lasing transitions in lithiumlike and hydrogenlike ions were found for a variety of materials. Figure 11.4 shows a schematic of a transient recombination X-ray laser scheme.

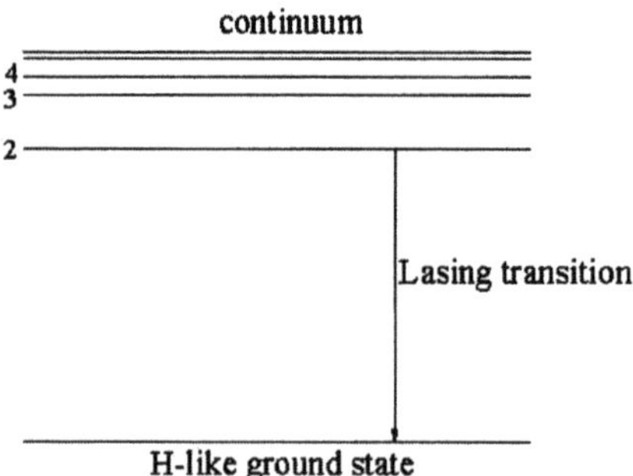

Fig. 11.4. Schematic of a transient recombination X-ray laser scheme

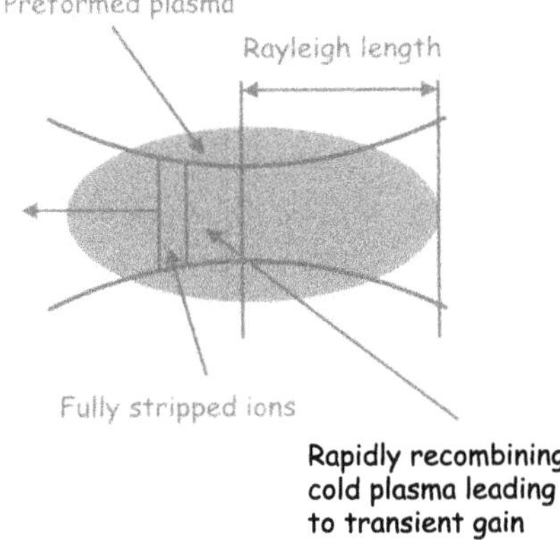

Fig. 11.5. Schematic view of OFI X-ray laser scheme

For density sufficiently low to avoid depletion of the driving laser pulse by absorption or ionization-induced defocusing (see Sect. 11.2), long-length plasma (over several Rayleigh length) with a high degree of ionization can be produced at a low input energy by optical-field ionization using confocal geometry (a schematic view of the experimental setup is given in Fig. 11.5).

As an example, a 100-fs Ti-sapphire with energy in the 100-mJ range focused at an intensity around $10^{17}\,\mathrm{W/cm^2}$ (w_0 =25 μm) can produce fully stripped lithium ions over 5 mm. Figure 11.6 shows the ionization threshold intensity as a function of the ionization potential calculated with a barrier-suppression ionization (BSI) model [153], s for heliumlike ions (solid square) and fully stripped ions (open circles) for materials from helium ($Z = 2$) to silicone ($Z = 14$).

Figure 11.7 gives the threshold ionization intensity as a function of the lasing wavelength for $2 < Z < 14$ calculated by the BSI model for Li-like schemes (open triangles) and hydrogenlike schemes (full squares). In Fig. 11.7, we have assumed that the wavelength of the lasing transition scales as Z^{-2}, where Z is the atomic number of the lasant material, as expected for recombination schemes.

On the other hand, low energy requirements allow the development of high repetition rate (10 Hz–1 kHz) X-ray lasers of particular interest for pump/probe experiments such as photoelectron spectroscopy [212] or for soft X-ray projection lithography (also called EUV lithography) [213].

Lasing can occur down to the ground state of the ion which allows achieving short lasing wavelengths, compared to lasing between excited states. However, the ground state must be completely empty because the population of the upper level of the transition can never be important due to the fast radiative decay down to the ground state. A small population of the ground state leads to a dramatic decrease of the gain of the transition [208]. The population of the ground state of the lasing ions as well as the required ion-

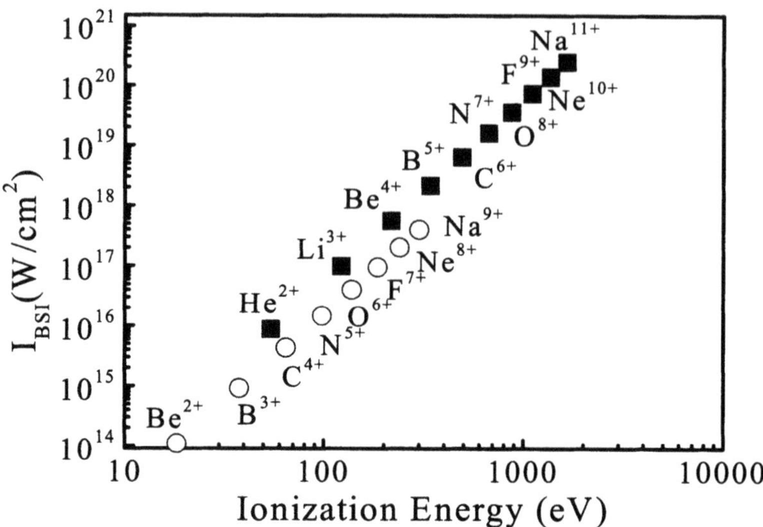

Fig. 11.6. Ionization threshold intensity versus ionization potential given by the BSI model for heliumlike ions (*solid squares*), and fully stripped ions (*open circles*) for materials from helium ($Z = 2$) to silicone ($Z = 14$)

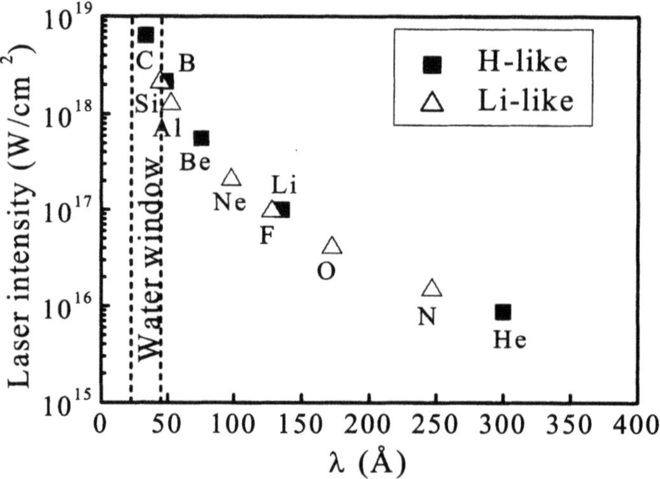

Fig. 11.7. Threshold ionization intensity versus lasing wavelength for $2 < Z < 14$ calculated by the BSI model for Li-like schemes (*open triangles*) and hydrogenlike schemes (*full squares*)

ization state in the focal volume can be precisely and easily controlled by monitoring the laser intensity because ionization depends strongly on intensity [214,215]. Moreover, no cooling is required in the OFI scheme, in comparison with conventional X-ray laser, since very low electron temperatures can be achieved in optical-field-produced plasma with a linearly polarized laser which is favorable for rapid recombination. For low gas density ($\sim 10^{17}\,\mathrm{cm}^{-3}$) and/or moderate laser intensity ($\sim 10^{17}\,\mathrm{W/cm}^2$), numerous preliminary experimental works performed on low (helium) and intermediate (Ne) Z material [216–221] have demonstrated that electron temperatures of a few eV (lower than the ionization potential) can be achieved. Such low temperatures correspond to the residual energy coming from the ionization process itself and depend on the phase of the electric field (for a linearly polarized wave) of the laser at the moment of ionization [222]. Figure 11.8 shows the energy of electrons resulting from the ionization of argon by a linearly polarized 130-fs Ti-sapphire laser ($\lambda = 0.8\,\mu\mathrm{m}$) as a function of the maximal intensity calculated with the Ammosov–Delone–Kraïnov (ADK) tunnel-ionization model [223].

The residual energy of the electrons increases with wavelength and the intensity of the laser [224] and with gas pressure. Blyth and co-workers [218] demonstrated with a KrF laser ($\lambda = 0.28\,\mu\mathrm{m}$) that for an intensity of $10^{18}\,\mathrm{W/cm}^2$ and pulse duration of 350 fs at intermediate density ($\sim 10^{18}\,\mathrm{cm}^{-3}$), the electron temperature is determined by ionization-induced heating (also called above-threshold ionization (ATI)) and by nonlinear inverse bremsstrahlung heating, whereas in the limit of high density ($> 10^{20}\,\mathrm{cm}^{-3}$), the plasma temperature is driven by strong inverse bremsstrahlung and stimulated Raman scattering heating, imposing a density and/or

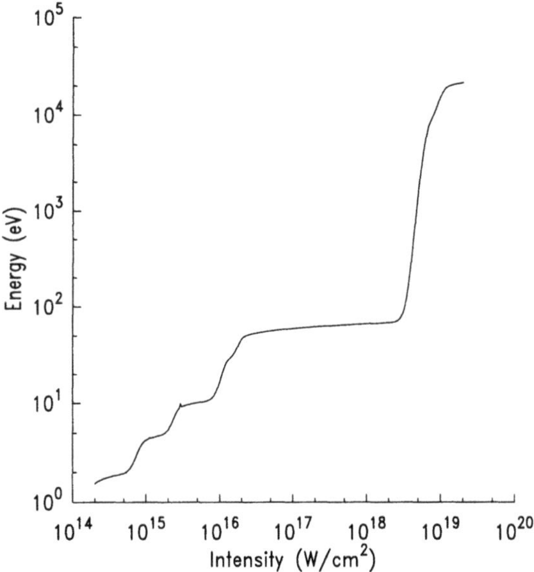

Fig. 11.8. Energy of electrons resulting from the ionization of nitrogen by a linearly polarized 130-fs Ti-sapphire laser ($\lambda = 0.8\,\mu\mathrm{m}$) as a function of the maximal intensity, calculated with the ADK model

intensity limit for the production of cool plasma, and thus a limit on the wavelength achievable with OFI X-ray laser schemes. Limitation of OFI schemes due to parametric heating was already mentioned by Amendt and co-workers [208] who found, using a particle-in-cell (PIC) code, that extending the scheme to wavelengths around 50 Å (e.g., Li-like Al at 52 Å) would not be possible because it would require laser intensity around 10^{18} W/cm^2 and the plasma temperature would then reach 1 keV for a 50-fs pulse. Finally, Glover and co-workers [220, 221] showed that the initial photoelectron distribution resulting from optical-field ionization of helium and neon is strongly non-Maxwellian and that subsequent recombination kinetics should then proceed more slowly than predicted for a Maxwell distribution of electrons.

The feasibility of OFI X-ray lasing was first demonstrated in H-like lithium on the Lyman α transition ($2p \rightarrow 1$ s) at 13.5 nm at the RIKEN laboratory in Japan [225].

A 20-ns KrF laser pulse was line focused onto a flat Li target at an intensity optimized to create a singly ionized plasma column ($10 \sim 10^9$ W/cm^2) with an electron temperature of 1.5 eV. A 50-mJ, 500-fs KrF was then focused in the plasma column about 700 ns after the nanosecond pulse, producing fully stripped Li ions and an electron density of 2×10^{17} cm^{-3}.

A small-signal gain coefficient of 20 cm^{-1} was obtained on a maximum length of 2 mm, giving a gain–length product of 4. Attempts to increase the gain–length product by increasing the plasma length beyond 2 mm were unsuccessful.

Experiments carried out at Berkeley/Livermore [226] and Princeton [227] showed the same limitation, although experimental conditions were slightly different. A possible explanation of this limitation is ionization-induced defocusing of the driving pulse. To prevent this effect, Korobkin and co-workers at Princeton [228] successfully used a 5mm long LiF microcapillary.

A low power, 2-Hz Nd:Yag laser (100-mJ, 5-ns) was focused on the entrance of a microcapillary. After a few hundred nanoseconds delay, a 0.3-ps KrF laser was focused with a maximal intensity of 2×10^{17} W/cm^2 onto the plasma at the entrance of the microcapillary. A study of the intensity of the LiIII 13.5-nm line as a function of microcapillary length demonstrated a small-signal gain coefficient of 11 cm^{-1}, i.e., a gain–length product of 5.5 for the 5-mm maximum lasing length. Until now, no other lasing material was found.

Another approach to tabletop size X-ray laser is to populate the upper level of the lasing transition by collisions between the lasing ions in their ground state and energetic electrons produced by optical-field ionization of the lasant material (a noble gas) with a circularly polarized, short-pulse, high-intensity laser; the residual energy of electrons produced with a circularly polarized laser pulse is higher than that of a linearly polarized one, as shown by Corkum and co-workers [222]. In this collisional scheme, proposed and demonstrated experimentally by Lemoff and co-workers [229, 230], lasing is

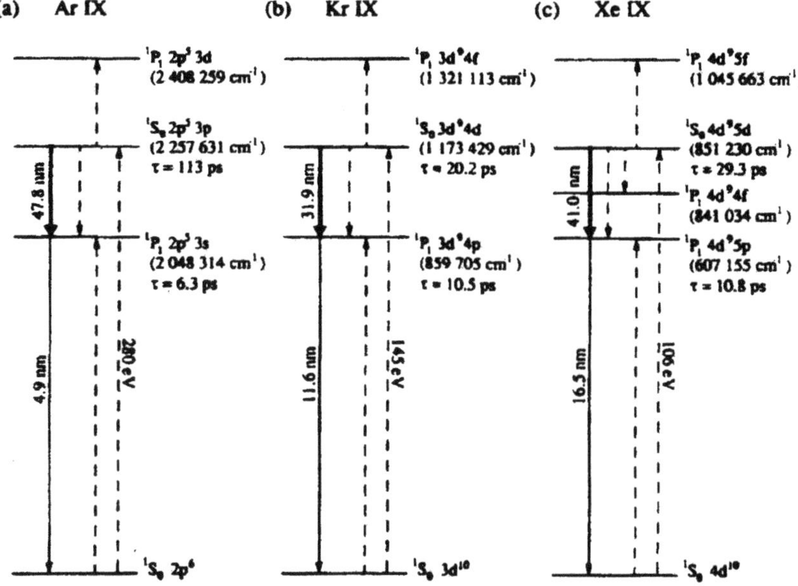

Fig. 11.9. Energy-level diagram of **(a)** Ar^{IX} (Ne-like), **(b)** Kr^{IX} (Ni-like), and **(c)** Xe^{IX} (Pd-like) [229]. Lasing transition is indicated by a thick arrow

between two excited states since a transition to the ground state is then optically forbidden. Figure 11.9 shows the energy-level diagram of (a) Ar^{IX} (Ne-like), (b) Kr^{IX} (Ni-like), and Xe^{IX} (Pd-like) [229].

The dashed lines indicate the most significant electron-induced transitions, and the solid lines give radiative transitions. The lasing wavelength is 47.8 nm in Ar^{IX}, 31.9 nm in Kr^{IX}, and 41.8 nm in Xe^{IX}. The experiment was carried out with a 10-Hz powerful Ti-sapphire laser (70-mJ, 40-fs) focused in a cell filled with low pressure (12 Torr) xenon. The maximal intensity was higher than 3×10^{16} W/cm^2 and produced Xe^{IX} ions on more than 7 mm. A study of the intensity of the 41.8-nm line with the length of the lasing medium showed a small-signal gain coefficient of 13 cm^{-1} and a gain–length product of 11. Despite its important gain–length product, the energy of the lasing transition is relatively low because it is between excited states. Moreover, the scheme seems to be quite difficult to extrapolate to shorter wavelengths because of the rapidly increasing intensity required to further ionize noble gas atoms and the difficulty in achieving the required intensity over a long distance.

A third approach to X-ray lasing using a high-intensity, short-pulse driving laser is the inner shell photoionization (ISPI) scheme. In this scheme, lasing is obtained by inner shell photoionization of the lasant material with an incoherent hard X-ray source produced by a powerful driving laser focused onto a solid target close to the lasant material. The inversion of population occurs between electron vacancies in the K shell (upper state) and L shell (lower state). This scheme requires more energy than the OFI scheme but can provide lasing wavelengths ($\lambda = 1.5\,\text{nm}$ for the K_α transition in Ne) which cannot be reached using OFI schemes because of parametric heating of the plasma due to high-intensity requirements. Such short wavelengths can neither be achieved with conventional collisional excitation schemes. This capability of ISPI X-ray lasers to generate coherent radiation with wavelengths as short as 0.5 nm makes them of particular interest for applications, e.g., in time-resolved crystallography [231].

The use of an X-ray source from laser-produced plasma as an optical pump to excite X-ray laser transitions is one of the earliest concepts proposed for creating X-ray lasers. This was first proposed by Duguay and Rentzepis in 1967 [232]. Soft X-ray-pumped inner shell photoionization lasers in the UV and the VUV range were demonstrated in the early 1980s by Silfvast and co-workers in cadmium vapor ($\lambda_{\text{emission}} = 442\,\text{nm}$, $325\,\text{nm}$) [233] and by Kapteyn and co-workers [234]. In the latter case, the laser transition ($\lambda_{\text{emission}} = 109\,\text{nm}$) was excited by Auger decay following photoionization of xenon. Demonstrations of this scheme were later done in krypton ($\lambda_{\text{emission}} = 91\,\text{nm}$) [235] and in zinc vapor at 127, 131, and 132 nm by Walker and co-workers [236]. Observations of gain saturation of the Xe^{III} 109-nm laser was also reported by Sher and co-workers [237] using traveling-wave laser-produced plasma excitation. Despite the progress made in demonstrating ISPI UV and VUV-lasers, the ISPI X-ray laser scheme has never been demonstrated because very bright X-ray pump sources and therefore a high-intensity driver are required to compensate for the rapid depletion of the upper state by nonradiative Auger decay. Typical Auger decay rates are around 10^{14}–$10^{15}\,\text{s}^{-1}$. One solution to this problem proposed by Mani [238] and co-workers and by Harris and Young [239] is to find an atomic level that is metastable to Auger decay. Thus, the radiative branching ratio and consequently the gain cross section are large. Unfortunately, this solution applies only for lasing wavelengths greater than 20 nm. The X-ray pumping source must also have a very fast rising edge to populate the upper state of the lasing transition faster than the lower state is populated by electron collisional ionization of neutral atoms. Collisional ionization can rapidly destroy the gain by simultaneously populating the lower state of the transition and reducing the neutral population which feeds the upper state.

With the development of high-intensity, short-pulse lasers, bright X-ray sources with ultrafast rising edges are now available. Observations of X-ray emission in the keV range with picosecond duration and conversion efficiencies of laser energy to X-rays of around 0.1% (in a 2π solid angle) were reported by several groups around the world [240–254]. Murnane and co-workers [255] and Gordon and co-workers [256] showed that X-ray emission can be significantly enhanced by using grating targets instead of flat targets. Conversion

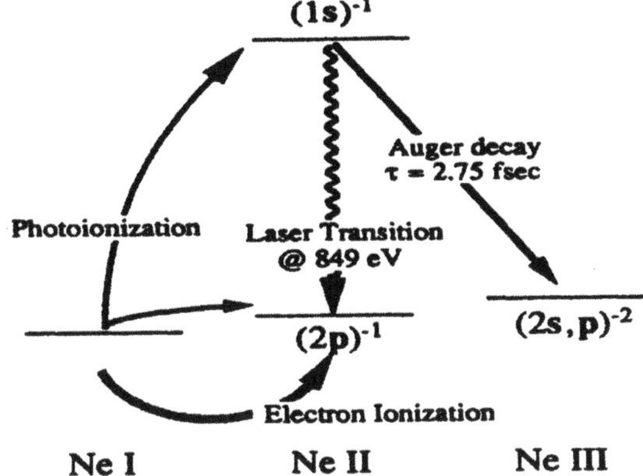

Fig. 11.10. Simplified energy level diagram of lasers at the Ne K_α transition at 1.5 nm [257]

efficiencies greater than 1% were obtained in the keV range by this method. These important advances in the production of high-brightness ultrashort-pulse X-ray sources led to new investigations of ISPI X-ray laser schemes, first by Kapteyn [257] and more recently by Borovsky and Mokrov [258] and by Li and co-workers [259]. Kapteyn considered the neon K_α transition at 1.5 nm (see Fig. 11.10).

He calculated that a small-signal gain coefficient of $10\,\mathrm{cm}^{-1}$ could be achieved, despite a short gain duration ($\sim$ 50 fs), for a neon density of $10^{20}\,\mathrm{cm}^{-3}$ diluted in solid hydrogen at density of $2 \times 10^{22}\,\mathrm{cm}^{-3}$ using a 200-TW (10-J, 50-fs) Ti-sapphire laser as a driving pulse. Hydrogen is used here as a moderator to limit electron collisional ionization of neutral atoms that populate the lower state of the lasing transition. A schematic view of the proposed experimental setup is shown in Fig. 11.11. The radiating plasma and the lasant medium are on both sides of a metallic filter. It is used to block photons with energy below the K-shell binding energy because they would contribute to populating the L shell, i.e., the lower state of the lasing transition.

The point of the paper by Borovsky and Mokrov [258] is that developing a laser on inner shell transitions is problematic due to self-limitation of the transition and to the absorption of the lasing line by the higher atomic shells, particularly by the outermost shell.

Calculations on the ISPI soft X-ray laser schemes proposed by Duguay in sodium vapor were also recently carried out by Li and co-workers [259]. The authors studied the influence of the rise time of the X-ray pulse on the maximal gain of the lasing transition. They concluded that the rise time must be as short as 1 ps and, consequently, that the high-intensity driving laser must have a contrast better than 10^5 to prevent heating of the target used to

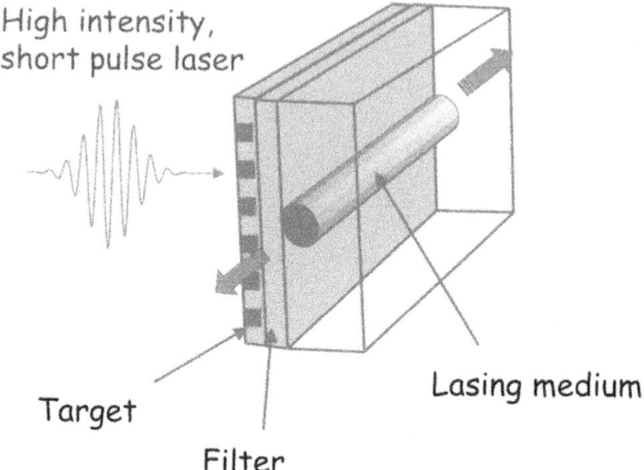

Fig. 11.11. Schematic of the experimental setup of an ISPI X-ray laser. A grating target is irradiated by a high-intensity, short-pulse laser that produced hard X-rays with high efficiency. Soft X-rays are blocked by a metallic filter. The lasing material is pumped by hard X-rays

generate the X-ray burst by a pedestral. Such a pedestral coming from the amplified spontaneous emission would lead to the generation of undesirable long-duration, slow-rise time, soft X-rays [246, 247].

11.3 Harmonic Excitation

With the rapid and recent advance in the technology of short-pulse, high-power lasers, many experimental and theoretical works on high-order harmonic generation of intense laser radiation interacting with gases [260–272], clusters [273], and solids [274–277] were carried out.

11.3.1 High-Order Harmonic Generation in Gases

Harmonics produced in low-pressure gases by atoms (or molecules) are emitted at odd-integer orders of the fundamental laser frequency along the propagation axis in the forward direction. They are observed for moderate laser intensities ($I \leq 10^{15}$ W/cm^2). A typical harmonics spectrum can be schematically divided into three regions, as can be seen in Fig. 11.12.

The first region of the spectrum corresponds to the low-order harmonics whose efficiency rapidly drops with the harmonic order. The second region is the plateau where harmonics efficiency is the same for all orders. The plateau extends up to a cutoff frequency corresponding to a maximum energy $E_{\max} = I_{\mathrm{p}} + 3 \times U_{\mathrm{p}}$, where I_{p} is the ionization potential of the gas and U_{p} is the ponderomotive energy [267, 278, 279]. In the third region, beyond

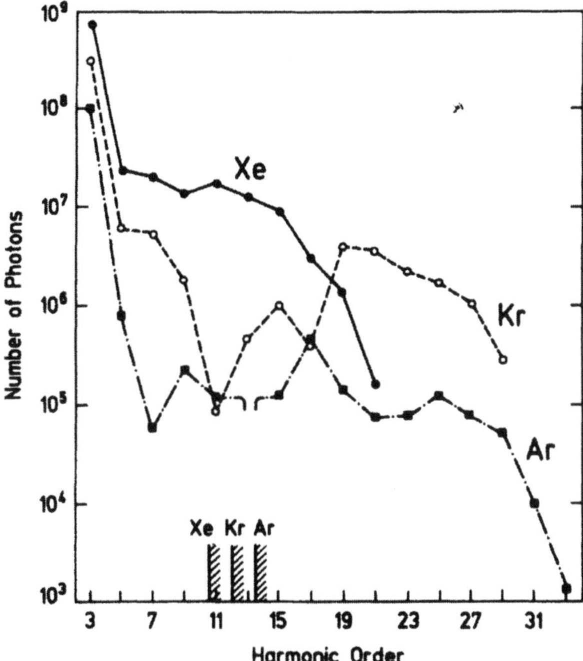

Fig. 11.12. Harmonic spectra obtained in Ar, Kr, and Xe with a Nd:Yag laser [260]

the cutoff, the harmonics signal rapidly drops to zero. The water window was recently reached with the 179th, and the 297th harmonic of a Ti-sapphire laser ($\lambda = 0.8\,\mu\text{m}$) [271, 272]. Spatial [280, 281] and temporal [282–284] coherence as well as focusing properties [285] and duration [286–289] of harmonics generated in gases were also extensively studied in the past few years, and harmonic radiations are now in use to probe the density of overcritical plasmas created by the interaction of short-pulse, high-intensity lasers with solid targets [290]. Harmonics generation beyond the cutoff frequency was also reported in several experiments carried out with Kr:F lasers with intensity up to $5 \times 10^{17}\,\text{W/cm}^2$ [265, 266, 291] and was attributed to emission from the low-charge-state ($Z = 1, 2$) ions of an underdense laser-produced plasma. Harmonics from a KrF laser emitted by the ions of a preformed highly ionized lithium fluoride plasma were also reported by Krushelnick and co-workers [292]. The ion emissions found were significantly weaker than those from atoms; the nonlinear polarization of the latter was higher. On the other hand, harmonic generation in rare gas clusters demonstrated somewhat higher efficiency in the cutoff region than dilute atoms [272].

11.3.2 Harmonic Generation in Plasmas

In underdense plasmas or ionizing gases, the propagation of laser pulses with relativistic intensity ($I \geq 10^{18}\,\mathrm{W/cm^2}$) can also lead to harmonics generation by electrons that undergo nonlinear motion due to their relativistic mass increase as their quiver velocity approaches the speed of light. Harmonics can either be emitted coherently by laser-driven nonlinear plasma currents or incoherently by nonlinear Thomson scattering.

In optical-field-ionized gases, the generation of a second harmonic resulting from the interaction of a high-intensity, short-pulse laser with an ionizing gas was observed by several groups [294–297]. In the early experiment of Liu and co-workers [294], a second harmonic was generated in the forward direction by focusing a 1-ps Nd:glass laser at moderate intensity ($I \leq 10^{16}\,\mathrm{W/cm^2}$) in a gas cell filled with hydrogen at low pressure ($P < 10\,\mathrm{Torr}$). The generation of a second harmonic was observed with both linearly and circularly polarized light, as expected in a medium with transverse density gradients. Furthermore, the harmonic signal, it was found, scales with the maximal laser intensity as $I^{3/2}$, once the intensity was larger than the ionization saturation threshold (i.e., the intensity for which the medium is fully ionized). Such scaling is characteristic of an increase in the ionized volume for a Gaussian beam and suggests that the second harmonic was generated on the edges of the ionized volume, i.e, in the gradient region. On the other hand, the study of the third-harmonic power with laser intensity, light polarization and gas pressure demonstrated that the third harmonic was generated by neutral gas in the rising edge and the far wings of the laser pulse. Measurements of the third-harmonic power generated in a preformed hydrogen plasma (i.e., a plasma composed only of free electrons and bare nuclei) showed that the conversion efficiency was at least five orders of magnitude less than that in a neutral gas. The conversion efficiency found was consistent with the value given by the relativistic cold fluid model developed by Esarey and co-workers [293].

Second-harmonic generation was also observed in experiments on relativistic self-channeling of high-power ($P \geq 10\,\mathrm{TW}$) laser pulses in optical-field-ionized gases [295–297] and preformed plasmas [298]. Emission was observed in the near forward direction [295] at 45° from the backward direction [296] and also at 90° from the laser propagation axis [297]. In that case, density gradients responsible for second-harmonic generation were created by optical-field ionization of the gas and/or the ponderomotive force associated with high-power, self-focused laser pulses. In the experiment by Malka and co-workers [295], spectra of second harmonics generated in the forward direction exhibited Stokes and anti-Stokes sidebands ($2\omega \pm 2n \times \omega_\mathrm{p}$) due to the scattering of second harmonics on a large-amplitude plasma wave generated by forward Raman scattering instability. Figure 11.13 shows spectra obtained for two different electron densities.

On the other hand, Krushelnick and co-workers [296] observed redshifts on second-harmonic spectra measured at 45° from the backward direction. They

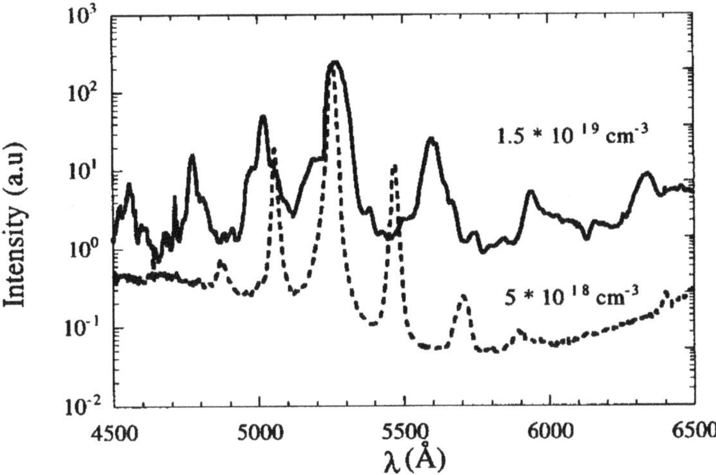

Fig. 11.13. Spectra of the second harmonic of a Nd:glass laser with Stokes and anti-Stokes sidebands, obtained for two different electron densities [295]

attributed these shifts to the frequency conversion of Raman sidescattered laser frequency, i.e., $(\omega - \omega_p) + (\omega - \omega_p) \rightarrow 2\omega - 2\omega_p$, in the density gradient due to the ponderomotive force.

Second- and third-harmonics generation by nonlinear Thomson scattering were also detected with linearly polarized laser pulses undergoing relativistic self-channeling in a high-pressure, optical-field-ionized helium gas jet [298]. The discrimination between Thomson scattered harmonics and plasma current-driven harmonics was identified by studying the angular distribution of second- and third-harmonic radiation in a plane perpendicular to the laser propagation axis because harmonics generated by nonlinear Thomson scattering, it is predicted, have typical angular patterns [34,35,299]. The angular pattern of a third-harmonic radiation is shown in Fig. 11.14. Furthermore, the harmonic signal found scaled linearly with the helium backing pressure, as expected for an incoherent emission process.

In addition to harmonic generation by gases and underdense plasmas, there is renewed interest in generating high-order harmonics by short-pulse, high-intensity lasers interacting with solid targets [274–277]. In that case, the generation results from large-amplitude, laser-driven electron oscillations in a density gradient steepened by the ponderomotive force. This mechanism was demonstrated in the early 1980s by Carman and co-workers [300,301], using a CO_2 laser ($\lambda = 10.6\,\mu m$) focused at an intensity around $10^{16}\,W/cm^2$ on a solid target. Both odd- and even-order harmonics were then produced and up to the 46th was observed. Recently, one-dimensional (1-D) PIC simulations from Gibbon [302] showed that a high-intensity ($I\lambda^2 > 10^{19}\,W{\times}cm^{-2}{\times}\mu m^2$), p-polarized pulse obliquely incident on a step-like density profile with $n_e = 10 \times n_c$, produced up to 60 harmonics with power conversion efficiency of

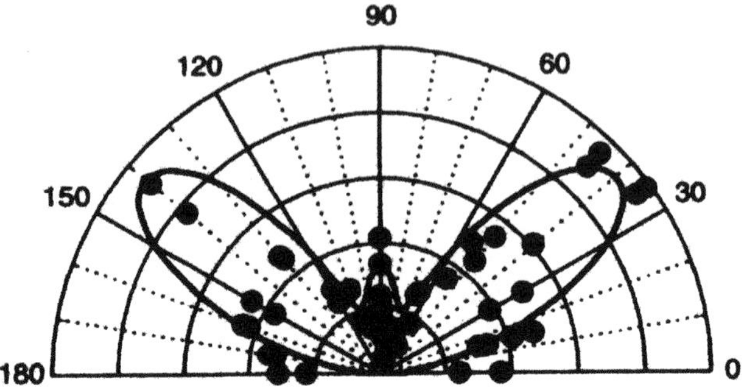

Fig. 11.14. Angular pattern of the third harmonic of a Nd:glass laser generated by nonlinear Thomson scattering. The solid curve is the theoretical prediction [298]

10^{-6} in the 60th harmonic. No cutoff was observed in the spectrum, unlike previous results obtained by Carman and co-workers [300].

According to Gibbon's numerical results, Norreys and co-workers [276] observed up to the 75th harmonic of the VULCAN Nd:glass laser system [187], focused at 10^{19} W/cm^2. Energy conversion efficiencies were estimated at between 10^{-4} and 10^{-6}. The harmonic emission found was isotropic, and no difference was observed between s- and p-polarization. The last point is in opposition to the numerical results of Gibbon who, however, emphasized that Rayleigh–Taylor-like instability at the critical layer, observed in 2-D PIC simulations [134] and due to the interaction of a high-intensity short-pulse laser with a preformed plasma, would scatter harmonics over a broad angular range relative to the specular direction and would blur the distinction between s- and p-polarized light.

11.4 Generation of Intense Electrostatic Fields and Acceleration of Electrons

When a short-pulse laser propagates through an underdense plasma, a large-amplitude Langmuir wave is excited in the wake of the laser pulse by the ponderomotive force associated with the temporal profile of the pulse. For tightly focused pulses ($k_{\mathrm{p}} \times w_0 \leq 1$, where k_{p} and w_0 are the plasma wavevector and the beam size at the waist, respectively), both components of the ponderomotive force, longitudinal and radial, generate a density perturbation, whereas in loosely focusing geometry ($k_{\mathrm{p}} \times w_0 \gg 1$), only a longitudinal electron plasma wave (EPW) is generated [303, 304]. The amplitude of the wave is maximum when $\omega_{\mathrm{p}} \times \tau \sim 1$, where τ is the pulse duration and ω_{p} is the plasma frequency. The corresponding maximum electric field generated at resonance is given by E_z (GV/m)$= 1.35 \times 10^{-18} \times I_{\max}$ [W/cm^2] $\times$ (λ [μm])$^2/\tau$ [ps], where $I_{\max}$ and

λ are the maximal intensity and the wavelength of the laser, respectively. The relative longitudinal perturbation of electron density is given by $\delta z = E_z/E_0$ where $E_0 = mc\omega_p/e$. The ratio between the radial component of the electrostatic field and its longitudinal component scales as $E_r/E_z = \sqrt{2}\lambda_p/\pi w_0$. Electrostatic fields of tens of gigavolts per meter can then be produced.

Laser-induced wakefields were first observed as coherent far-infrared radiation from laser-produced plasma by Hamster and co-workers [305, 306], but the first spatiotemporal characterization of a density perturbation in the wake of an ultrashort laser pulse was done by Marquès and co-workers [307] and Siders and co-workers [308, 309], using a spectral interferometric technique [310]. Both experiments were carried out in the two-dimensional regime ($k_p \times w_0 \leq 1$), i.e., the radial component of the electrostatic field was greater than the longitudinal component. In both cases, a Ti-sapphire ($\lambda = 0.8\,\mu$m) laser with a 100-fs pulse duration and 0.3-TW power was focused at maximal intensity around $3 \times 10^{17}\,\mathrm{W/cm^2}$ in a cell filled with a low pressure gas, creating a plasma whose density satisfied the resonance condition $\omega_p \times \tau \sim 1$. Marquès and co-workers measured a relative density perturbation between 30 and 100%, depending on the density. The details of the spectral interferometric technique are given in [310] and in references therein.

On the other hand, Siders and co-workers measured a large-amplitude plasma perturbation ($\delta n \sim n_e$) corresponding to a longitudinal electric field of $\sim$ 10 GV/m, in agreement with results of a two-dimensional relativistic model of laser wakefield excitation [309]. Damping of radial electron density oscillations and a nonlinear plasma frequency increase predicted by Dawson [313] were also measured by Marquès and co-workers [311, 312], using the same spectral interferometric technique that they previously used [310]. The plasma frequency upshift may be due to several effects. In hot plasmas (with $T_e \neq 0$), the frequency is upshifted according to the Bohm and Gross relation of dispersion. A blueshift may also be due to a geometric effect [313]. In cylindrical geometry, the radial motion of electrons away from the axis of symmetry produces a charge density on-axis larger than in planar symmetry and generates a stronger restoring electric field which in turn leads to an up-shift of the plasma frequency. The damping of the electron plasma wave (EPW) was attributed by Marquès and co-authors to dephasing of a fraction of the electrons with the EPW. This phase mismatch is introduced by the two-step radial density gradient arising from optical-field ionization of helium. Electrons close to the He^{2+}/He^{+} interface radially accelerated by the ponderomotive force feel either the He^{2+}/He^{+} gradient or the He^{+}/vacuum gradient and go back to the plasma after a few oscillations. They are then out of phase with the EPW and destroy the oscillation. It has been observed that damping increases with decreasing pressure. This tends to confirm the importance of the ponderomotive force in the damping mechanism since the electron excursion increases as the space-charge field decreases.

The generation of very strong electric fields in the wake of high-intensity, short-pulse laser propagating in a low-density plasma makes it an attractive way for accelerating electrons. The laser wakefield accelerator (LWFA) was proposed 20 years ago by Tajima and Dawson [314]. This acceleration scheme presents three major interests. First, the phase velocity of the Langmuir wave is equal to the group velocity of the laser in the plasma. For this reason, the plasma wave excited in this way is also called a "relativistic" plasma wave. Second, it is not affected by saturation as relativistic detuning [315] or modulational instability [316], as the beat-wave is. Third, the experimental conditions to excite a plasma wave are much easier to fulfill than in the beat-wave experiments [315,317–322] where the EPW is resonantly excited by two temporally beating laser pulses ($\Delta\omega = \omega_\mathrm{p}$) that must overlap in space and time.

Laser Wake-Field Accelerator was recently demonstrated by Amiranoff and co-workers [323] in a regime where the plasma wave was mainly excited in the radial direction. The transverse component of the electric field was therefore stronger than longitudinal one. The cw electron beam from a Van de Graaff accelerator with a 3-MeV energy and 300-μA current was injected in the wake of a 400-fs, 3.5-TW, 1.057-μm laser focused to a maximal intensity of $4 \times 10^{17}\,\mathrm{W/cm^2}$ into a gas cell filled with helium. The gas pressure was adjusted to obtain the maximum plasma-wave amplitude. The maximum longitudinal electric field was estimated at 1.5 GV/m, and the maximum energy gain was measured at 1.6 MeV. The plasma-wave lifetime inferred from the number of accelerated electrons (coming from a cw beam) was 1 ps. This experiment also emphasized the difficulties inherent in the injection of electrons in electric field with an important transverse component. Depending on their phase, electrons are focused or defocused when they enter the plasma wave. The defocused electrons are expelled radially before they reach the laser beam waist, i.e., the region where the longitudinal accelerating electric field is maximum. The focused electrons remain trapped and those with a suitable phase can gain the maximum energy available $\Delta W = \pi e z_0 E_z$, where z_0 is the Rayleigh length. It was also pointed out that the emittance of the electron beam plays a crucial role in the fraction of accelerated electrons; the total number of electrons accelerated decreases dramatically with increasing emittance. Finally, it was shown that in such experimental conditions ($E_r > E_z$), the electron trapping distance must not exceed the dephasing length which is the length necessary for the electrons to be out of phase with the EPW. Otherwise, most of the electrons are expelled from the EPW during their transit through the wave, and only a small fraction is accelerated throughout the plasma.

With a high-intensity laser, large-amplitude relativistic plasma waves can also be driven by stimulated Raman forward scattering (RFS) instability. Stimulated RFS instability occurs when a high-intensity laser pulse is propagated through a plasma whose period is much shorter than the pulse duration.

The strong coupling between the laser and the plasma gives rise to local modulations in the index of refraction at the plasma wavelength which in turn break the laser pulse into multiple pulselets having a duration corresponding to the plasma period. When the laser power is greater than the critical power for self-focusing, stimulated RFS instability can be enhanced by relativistic self-focusing [326–328]. High-density plasmas can support large acceleration fields ($\sim 100\,\mathrm{GV/m}$) before the onset of wave-breaking (which happens when the growth of the plasma wave is so important that the waveform is no longer sinusoidal but steepened so extremely that there are singularities in the plasma density), i.e., before the plasma wave reaches its maximum amplitude. The accelerating longitudinal electrostatic field predicted by one-dimensional cold-fluid theory is then given by $E_{WB} = [2(\gamma_\mathrm{p}-1)]^{1/2} \times E_0$ (the electric field of wave breaking), where $\gamma_\mathrm{p} = 1/(1 - v_\mathrm{p}^2/c^2)^{1/2}$ and $E_0 = mc\omega_\mathrm{p}e$. Typically, $\gamma_\mathrm{p} \approx \omega/\omega_\mathrm{p}$, e.g., $\gamma_\mathrm{p} \sim 10$ and $E_0 \sim 300\,\mathrm{GV/m}$ for $n_\mathrm{e} \sim 10^{19}\,\mathrm{cm}^{-3}$. In the wave-breaking limit, a fraction of the electrons from the plasma is trapped in the electron plasma wave and is continuously accelerated to relativistic energies. Even though the electron plasma wave grows from an instability and consequently its phase is unpredictable, this self-modulated mode provides a compact accelerator (often called "self-modulated" laser wakefield accelerator (SM-LWFA)) which does not require injecting electrons and is not diffraction limited as standard LWFA is, because relativistic self-channeling of the laser can be used to enhance the accelerating region well above the Rayleigh length. Moreover, SM-LWFA can produce an accelerating electric field much larger than the standard LWFA because of the high-density plasma used ($n_\mathrm{e} \sim 10^{19}\,\mathrm{cm}^{-3}$). On the other hand, the use of too high an electron density would lead to a reduction of the Lorentz factor ($\gamma_\mathrm{p} \sim \omega/\omega_\mathrm{p}$) and therefore to a decrease of the phase velocity of the plasma wave.

In the past 5 years, many experimental results on self-modulated LWFA have been obtained. Correlation of energetic electrons with an anti-Stokes sideband ($\omega + n \times \omega_\mathrm{p}$) in the laser spectrum transmitted through the plasma, characteristic of stimulated RFS instability, was first reported by Coverdale and co-workers [329]. A few years later, acceleration of externally injected electrons [330] and of self-trapped electrons (up to 100 MeV) [183, 331–335] was demonstrated. Temporal characterization of a self-modulated wakefield was also performed [181, 336].

The experimental demonstration of electron acceleration by a self-modulated wakefield was first performed by Nakajima and co-workers [330]. They injected 0.6-MeV electrons produced from a solid target irradiated by a long-pulse (200-ps), high-intensity ($> 10^{16}\,\mathrm{W/cm^2}$) laser into a 30 GeV/m longitudinal electrostatic field generated into an optical-field-ionized high-density helium gas jet by a 3-TW, 1-ps Nd:glass laser system (1.052 μm). Electrons with 18 MeV maximum energy were measured. They estimated that about 0.2% of the injected electrons have energy up to 5 MeV. Just after Nakajima and co-workers, Modena and co-workers [331, 332] reported experimental results on the acceleration of self-trapped electrons. This experiment was performed with the VULCAN laser system [187], which was focused at a maximal intensity of 6×10^{18} W/cm^2 onto a high-pressure helium gas jet. The number of accelerated electrons at a given energy as well as the maximum electron energy increased dramatically with plasma density. The number of electrons with energy above 20 MeV increased by a factor of two when the density increased from 1.0×10^{19} cm^{-3} to 1.5×10^{19} cm^{-3}. In the latter case, electrons with energy up to 44 MeV were observed. From this maximum energy, a maximal accelerating electric field of 100 GV/m was inferred. Wave-breaking was used as the explanation for the trapping and acceleration of background electrons by the EPW to the high energies measured. Simultaneous detection of the transmitted light spectrum showed that the Stokes ($\omega - n \times \omega_p$) and anti-Stokes ($\omega + n \times \omega_p$) sidebands which are clearly seen at low plasma density (a few times 10^{18} cm^{-3}) broaden and nearly disappear at higher density (1.5×10^{19}cm^{-3}) (see Fig. 11.15). Modena and co-workers attributed this broadening of the plasma satellites to a destruction of EPW coherence and interpreted it as a signature of wave-breaking. A typical spectrum obtained in the wave-breaking regime is shown in Fig. 11.16.

More recently, Ting and co-workers [333] and Moore and co-workers [334] observed electron energies up to 30 MeV for the same plasma conditions as Modena and co-workers but for a laser power 10 times lower (2.5 TW) and without any evidence of wave-breaking. They explained their results by preacceleration of background electrons before they are trapped by the wakefield. Electrons would be preaccelerated by a low phase velocity wave which would result from the beating between the forward intense laser wave and the backscattered electromagnetic wave reflected by stimulated backward

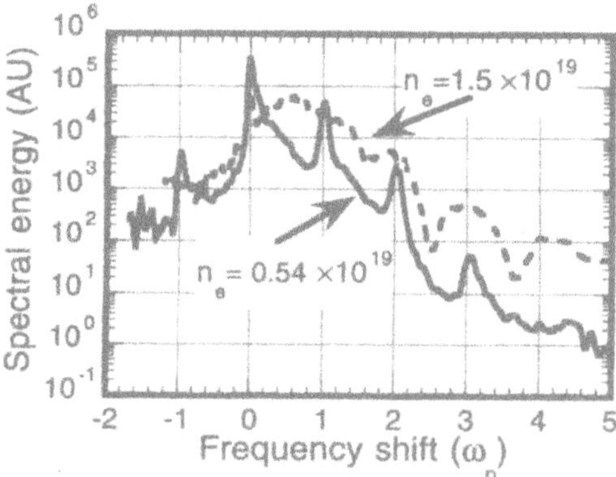

Fig. 11.15. Transmitted laser spectrum with Stokes and anti-Stokes sidebands due to RFS, measured for two different electron densities. One can clearly see the broadening of plasma satellites in the spectrum obtained at the highest density [331, 332]

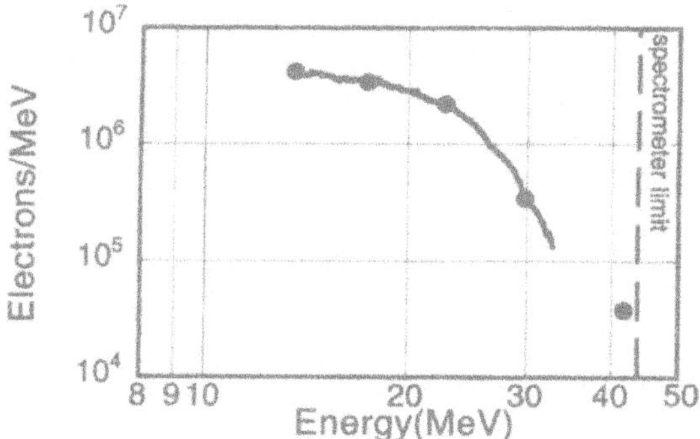

Fig. 11.16. Electron spectrum obtained for $n_e = 1.4 \times 10^{19}\,\mathrm{cm}^{-3}$ [332]

Raman scattering (BRS). Simultaneously with Modena and co-workers, Umstadter and co-workers [183] reported observing a relativistic, high-current, low-emittance, electron beam created in the relativistic self-channeling regime by focusing a 7.5-TW, 0.4-ps Nd:glass laser onto a high-pressure jet of helium (or argon) gas, producing a plasma with a $3 \times 10^{19}\,\mathrm{cm}^{-3}$ electron density and a relative density perturbation between 8 and 40%, giving a maximal accelerating field between 0.5 GV/m and 2 GeV/m. The number of electrons with energy corresponding to the peak of the distribution, i.e., 2 MeV, was measured as a function of the incident laser power for a fixed backing pressure. A dramatic increase of the number of 2-MeV electrons was observed once the laser power is higher than the critical power for self-focusing, as shown in Fig. 11.17.

At the highest power (7.5 TW), more than 10^9 electrons were accelerated. Images of the output electron beam showed a transverse emittance as low as 1

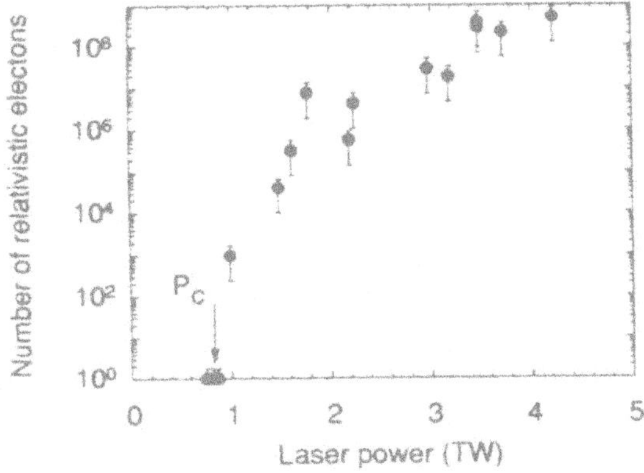

Fig. 11.17. Number of 2 MeV electrons as a function of laser power [183]

millimeter–milliradian. Such performances in terms of transverse emittance and number of electrons (>0.5 n_c) are both comparable to state-of-the-art current radio-frequency photoinjector linear accelerators. Further investigations by Wagner and co-workers [337] using the same experimental conditions clearly demonstrated correlations between the distance over which the laser beam remained relativistically self-guided on one hand, and the divergence of the outgoing electron beam as well as the energy of the electrons, on the other hand. It was found that the divergence of the electron beam decreases and the energy of the electrons increases when the self-channeling length increases.

Finally, temporal characterization of the self-modulated wakefield by Le Blanc and co-workers [336] using the same conditions as those of Umstadter and co-workers and Wagner and co-workers showed growth and decay rates consistent with stimulated RFS instability and Landau damping, respectively; the latter arises from the transfer of energy from the electron plasma wave to the self-trapped electrons. Similar results were obtained by Ting and co-workers [181] for comparable experimental conditions.

11.5 Generation of Superintense Magnetic Fields

In the past few years, it was demonstrated that high-intensity, short laser pulses interacting with underdense or overdense plasmas can accelerate electrons to ultrarelativistic energy (up to 100 MeV). This hot tail is produced either by wakefield or ponderomotive acceleration of background electrons. Intense electron beams copropagating with the laser pulse are then generated, and azimuthal magnetic fields of hundreds of MG have been observed in numerical simulations.

In a theoretical and numerical study relevant to the laser wakefield accelerator (LWFA) concept (see Sect. 11.4), Gorbunov and co-workers [338] showed that in a frame moving with a laser pulse, a quasi-static azimuthal magnetic field, with both a homogeneous component in the longitudinal direction and a $\lambda_p/2$ oscillating component, must exist in the wake of a high-intensity laser propagating in an underdense plasma. The homogeneous component found was due to the steady current created by the plasma wakefield, whereas the $\lambda_p/2$ oscillating component is a general feature of nonlinear plasma waves [339, 340]. This field is always negative and therefore should have a focusing effect on relativistic electrons injected in the wakefield. On the other hand, the magnetic field found increased with the laser spot size and the intensity in the weakly relativistic regime. For $a \geq 1$, the periodic structure disappears, and the field amplitude dramatically drops. No experimental evidence of such magnetic fields was obtained to date.

Three-dimensional PIC simulations by Pukhov and Meyer-ter-Vehn of the propagation of short laser pulses with relativistic intensity in near-critical plasma ($n_e \sim 0.4 \times n_c$) [341] show that intense electron jets, axially copropagating with the laser pulse, can generate azimuthal magnetic fields up to

100 MG. Such intense fields strongly modify the index of refraction of the plasma and therefore the laser propagation. Multiple light filaments resulting from current filamentation due to stimulated Raman scattering and to Weibel instability [342], it was observed, coalesce into a single one, and the laser intensity on-axis increases. The coalescence of light filaments is directly related to the magnetic collapse of current filaments. An azimuthal magnetic field surrounding a density channel created by the VULCAN laser [187] propagating in a slightly underdense preformed plasma ($n_e \sim 0.1 \times n_c$) was recently observed by Borghesi and co-workers [343]. The field magnitude was measured using Faraday rotation of a linearly polarized optical probe beam ($\lambda = 0.622\,\mu m$), propagating orthogonally to the main beam. Two magnetic field components with opposite sense of rotation were observed in the Faraday rotational patterns. One field component was detected in the outer region of the plasma. The other component, it was observed, remained confined within the channel wall during the expansion phase. The .sense of rotation found was the same as observed in previous measurements in longer pulse regimes [344, 345] and is consistent with fields generated by a thermoelectric mechanism (resulting from $\nabla T \times \nabla n_e \neq 0$ [346]); the inner field sense was consistent with a current of electrons accelerated in the forward direction by the laser. The amplitude of the inner component, it was estimated, was in the 1–10 MG range.

Several numerical studies on the fast igniter concept relevant to inertial confinement fusion (see Sect. 11.7) also predicted the existence of strong magnetic fields (~100 MG) in overdense plasma, generated by electrons from the critical layer accelerated in the forward direction to relativistic energy by the ponderomotive force associated with high-intensity laser pulses. In particular, Pukhov and Meyer-ter-Vehn [347] observed, in two-dimensional PIC simulations, magnetically confined electron jets with maximal current densities as high as $10^{13}\,\mathrm{A/cm^2}$ at a laser intensity of $10^{20}\,\mathrm{W/cm^2}$. The laser-driven electron jet, it was observed, breaks up into several filaments at an early time, when penetrating the overdense region, and then coalesces into a single filament later on. As will be discussed in Sect. 11.7, evidence of collimated relativistic electron jets was obtained in several experiments for laser intensity in the $10^{19}\,\mathrm{W/cm^2}$ range. However, magnetic fields inside the dense core were never explored because the Faraday rotational technique using a probe beam whose wavelength is in the visible or near-infrared range cannot be used due to refraction in the steep gradients present in the denser region of the plasma. Borghesi and co-workers [348] recently reported measurements of a magnetic field in plasma created by focusing the VULCAN laser system [187] at an intensity between 5 and $8 \times 10^{18}\,\mathrm{W/cm^2}$ onto a flat solid target. The field was measured in a region whose density was lower than $10^{20}\,\mathrm{cm^{-3}}$, using Faraday rotation of a probe beam at wavelength $\lambda = 0.622\,\mu m$. It was demonstrated that a MG transient azimuthal field with a typical lifetime of a few tens of picoseconds was generated. It was also observed that such fields modify the plasma expansion and confine it to a collimated plume. Two-dimensional magnetohydrodynamic (MHD) simulations have shown that the magnitude and the orientation of this field were consistent with the thermoelectric effect. Such magnetic confinement of a high-intensity laser-produced plasma was already inferred from ion-velocity measurements and time-integrated X-ray images by Bell and co-workers [349].

Another source of strong magnetic fields in plasmas is the so-called inverse Faraday effect [350]. In this case, the magnetic field is induced by the orbital motion of single electrons in a circularly polarized light wave. The macroscopic magnetization of the plasma results from superposition of all of the microscopic magnetic dipoles. The inverse Faraday effect in a plasma was demonstrated experimentally in the early 1970s by Deschamps and co-workers [351] in the microwave range. It was shown that in the low-intensity limit, the amplitude of the magnetic field was proportional to the intensity of the incident electromagnetic wave. With the high laser intensities currently achieved ($I \geq 10^{19}\,\mathrm{W/cm^2}$), magnetic fields as large as tens of MG can be produced, and there is a renewal in interest in generating strong magnetic fields by the inverse Faraday effect [352–359]. However, no experimental evidence of such huge fields was obtained to date. Generation of strong magnetic fields with high-intensity lasers was already considered in the early 1970s by Steiger and Wood [360] who calculated the field produced by a circularly polarized plane wave whose intensity was in the range 10^{17}–$10^{20}\,\mathrm{W/cm^2}$ propagating in a near-critical homogeneous plasma. They found that magnetic fields as high as 1 MG should be created at a laser intensity of $10^{18}\,\mathrm{W/cm^2}$ and a plasma density of $10^{20}\,\mathrm{cm^{-3}}$. More recently, Sheng and Meyer-ter-Vehn [359] considered the generation of magnetic fields by the inverse Faraday effect for relativistic intensities and in the presence of inhomogeneity in the transverse plasma density and laser intensity profiles. It was found that a magnetic field was produced by the circular motion of single electrons in the circularly polarized wave and also by the nonzero azimuthal currents due to density and intensity gradients. It was also shown that the generation of strong magnetic fields should give rise to a reduction of the critical power for relativistic self-focusing by a factor of $(1+\omega_\mathrm{p}^2/\omega^2)^{-1}$ and should also reduce electron cavitation due to the transverse ponderomotive force associated with the laser pulse undergoing self-focusing.

11.6 Interaction of Free Electrons with Ultrashort Laser Pulses

The interaction of light with free electrons gives rise to very well-known effects such as Thomson [361], Compton [362] or Breit–Wheeler scattering [363]. Thomson scattering is a process in which the energy of the incident photon is much smaller than the electron rest mass. Compton scattering occurs in the opposite case, i.e., the photon energy and the electron rest mass are of the same order of magnitude, and the electron recoils when interacting with the photon (theoretical studies of this phenomenon are described in Chaps. 4, 5, and 6). Breit–Wheeler scattering is a two-photon process that leads to electron–positron pair production. All of these scattering processes were first observed in the linear (nonrelativistic) regime of interaction, i.e., for $a^2 \ll 1$ (where a is the normalized vector potential). With the recent development of high-intensity short-pulse lasers that can produce electric fields as high as 3×10^{13} V/m, the scattering processes described above were studied in the nonlinear (relativistic) regime [298, 364–369].

Nonlinear Thomson scattering leads to the generation of harmonics with particular angular radiation patterns. Experimental evidence of this process was obtained via second- and third-harmonic detection [298]. The experiment is described in Sect. 11.3.

The transition between Thomson and Compton scattering processes is possible even though the photon energy is lower than the electron rest mass, provided that a large number of photons is simultaneously absorbed, i.e., on a timescale shorter than the typical electron–electron or electron–ion collision timescales. The transition between Thomson and Compton regimes was observed by Moore and co-workers [364] and Meyerhofer and co-workers [365] by studying the angular distribution of electrons created by optical-field ionization of neon gas at a very low density with a high-intensity ($I \sim 10^{18}$ W/cm^2), short-pulse ($\tau = 1$ ps) laser. They observed that the photoelectrons released from the first charged states, i.e., those that did not experience a high field, were ejected in the plane of polarization, as usually observed in the weak-field regime, whereas those coming from highly charged states, which feel the highest field strength, it was found, have nonzero velocity along the laser propagation axis.

Acceleration of free electrons to MeV energies in vacuum was observed by Malka and co-workers and was also attributed to Compton scattering [366]. Electrons with a few tens of keV energy generated at low-intensity (10^{12}–10^{13} W/cm^2) focused onto a solid plastic target were injected into the P102 laser beam focused in vacuum at an intensity of 10^{19} W/cm^2 ($a^2 \sim 9$). Side scattered electrons with energy up to 1 MeV were then detected in the plane of polarization of the high-intensity laser pulse in the forward direction.

Nonlinear Compton scattering was observed by Bula and co-workers [367] in an experiment carried out at the Standford Linear Accelerator Center (SLAC). They measured the energy of electrons scattered at the focus of a 1-TW, 1-ps, Nd:Yag laser counterpropagating with a 47-GeV electron beam. At the laser intensities achieved ($I \sim 10^{18}$ W/cm^2) and the wavelengths used (1.054 and 0.527 μm), the value of a^2 was equal to 0.8, and the photon energy in the rest frame of the electrons was 211 and 421 keV, respectively. Therefore, the laser field was strong enough to observe nonlinear effects, and the recoil of the scattered electrons was large enough to identify the scattering process as Compton scattering. In these conditions, simultaneous scattering of up to four photons was observed. The dependence of the scattered electron yield on the laser intensity was in quite good agreement with calculations developed in the 1960s by Nikishov and Ritus [370] and by Narozhnyi and co-workers [371]. An investigation of pair production at the focus of a high-power laser beam undergoing relativistic self-focusing in a plasma was also recently presented by Borovsky and co-workers [125].

In the same experiment, positrons resulting from pair production were also detected [368]. Positrons were created in a two-step process in which laser photons are first backscattered by the electron beam with GeV energies, and then each of them interacts with several incident laser photons to produce an electron–positron pair. The two-step reaction can then be written as

$$e + n \times \omega_0 \rightarrow e' + \omega \,,$$
$$w + n \times \omega_0 \rightarrow e + e^- \,,$$

where the first reaction corresponds to nonlinear Compton scattering and the second is the multiphoton Breit–Wheeler reaction [363]. The variation of positron rate with the laser field-strength parameter (defined in [364, 365, 367–369] as $a^2/2$) was well reproduced by a model describing the two-step Breit-Wheeler process.

Finally, as a last example of the interaction of free electrons with high-intensity laser pulses, let us cite the experiment of Leemans and co-workers [372, 373] who produced a short pulse (300 fs) of hard X-rays (30 keV) by 90° Thomson scattering of a 50-fs, 2-TW, 10-Hz Ti:sapphire laser on a 50-MeV electron beam. In this way, 10^5 X-ray photons with 30-keV energy were produced, leading to a conversion efficiency of 10^{-8}. The maximal spectral brightness found was also comparable to a synchrotron radiation source whose electron beams are in the range of 6–10 GeV.

11.7 Fast Igniter Scheme

The concept of fast ignition for inertial confinement fusion (ICF) was originally proposed in 1994 by Tabak and co-workers [374]. It is based on a pioneering numerical study by Wilks and co-workers [134]. In the fast igniter scheme, the fuel of a precompressed capsule is ignited by electrons accelerated to relativistic energy by the ponderomotive force associated with an ultrahigh intensity ($> 10^{19}$ W/cm^2) laser. Note that in this new approach to ICF, all of the effects described in this book are present: stimulated Raman scattering, wakefield and ponderomotive acceleration of electrons, relativistic self-focusing, relativistic and ponderomotive self-channeling, and MG magnetic field generation.

Using a two-dimensional electromagnetic particle-in-cell (PIC) code, Wilks and co-workers [134] demonstrated that the ponderomotive force associated with a high-intensity linearly polarized laser ($> 10^{18}\,\mathrm{W/cm^2}$) is so strong that the laser wave can penetrate an overdense plasma. First, the light penetrates a thin layer which is larger than the classical skin depth ($\sim c/\omega_\mathrm{p}$) due to the relativistic mass increase of the electron oscillating in the strong laser field. Such a modification of the classical skin depth is called self-induced transparency [375]. Second, in the interface between the layer and the unperturbed plasma, a high-amplitude electric field driven by the ponderomotive force is generated at a $2\omega_0$ frequency. This electric field accelerates electrons to very high energy ($\sim 1\,\mathrm{MeV}$) into the overcritical plasma with a high conversion efficiency of laser energy into kinetic energy ($\sim 30\%$). The departure of the electrons then creates a hole in the density profile which allows the light wave to penetrate deeper inside the plasma. It is so-called "hole boring." On the other hand, as the electrons are expelled, an electric field builds up that drags the ions, producing highly energetic ions. Strong quasi-static magnetic fields are also generated by the relativistic electron beam propagating inside the target, as discussed in Sect. 11.5.

With the rapid and continuous increase in laser performance, studies related to the fast igniter concept are now under way at a number of laboratories around the world and important progress in the basic understanding of the physical effects involved was made in the past 3 years. The generation and the transport of suprathermal electrons [376–384] were studied, as well as fast ions production [385, 386]. Plasma confinement due to the ponderomotive force [387, 388], relativistic and ponderomotive self-channeling of intense laser pulses in preformed underdense plasmas [389–392], as well as self-induced transparency through thin solid foils [393, 394] and hole boring in overcritical plasmas [395–400] were demonstrated. Evidence of very strong magnetic field generation was also obtained [343, 348].

During the very early phase in the fast igniter scheme, the high-intensity laser pulse must propagate through the underdense uncompressed plasma composed of material ablated onto the capsule containing the fuel. Propagation of a short-pulse, high-power laser in a preformed plasma with density $n_\mathrm{e} \sim 0.5 \times n_\mathrm{c}$, i.e., density of particular interest for the fast igniter scheme, was first examined by Young and Bolton [389]. They analyzed the temporal and spectral shape of the laser light transmitted through the plasma, using a frequency-resolved optical gating (FROG) diagnostic [401]. With this technique, they investigated the formation of a density channel in the plasma. They observed a redshift of the laser wavelength on the leading front of the pulse, as expected when the density channel forms and the electrons are expelled from the laser beam. On the trailing edge of the pulse, they measured a rapid blueshift with low-frequency modulations. These modulations were attributed to plasma oscillations in the low-density channel created by the laser. The rapid blueshift was explained by rapid collapse of the density channel behind the laser pulse. Simultaneously with Young and Bolton, Borghesi and coworkers [390] reported observations of relativistic self-channeling of the VUL-

CAN laser system [187] in a preformed plasma with density $n_e \sim 0.05 \times n_c$. In that case, images of the plasma obtained at the $2\omega_0$ frequency in a direction perpendicular to the propagation axis showed radial oscillations due to successive self-focusing and defocusing of the laser beam. PIC simulations in 3-D performed for plasma and laser parameters close to those of the experiment closely reproduced experimental observations and predicted the generation of relativistic electrons currents copropagating with the laser pulse and generating very strong (up to 100 MG) quasi-stationary toroidal magnetic fields. These fields were measured by Borghesi and co-workers [343]. Results of the experiment are presented in Sect. 11.5. Complementary information on the channel dynamics was obtained later by Fuchs and co-workers [391]. Using an interferometric technique, they studied the radial density profile and the transverse expansion of the channel created by the interaction of the P102 laser system [185] focused at an intensity of $5 \times 10^{18}\,\mathrm{W} \times \mathrm{cm}^{-2} \times \mu\mathrm{m}^2$ onto an underdense preformed plasma with initial density $n_e \sim 0.2 \times n_c$. They observed that the channel density varies dramatically with the laser intensity; the relative density perturbation $\Delta n/n_0$ (where n_0 is the background density) on-axis increased from 20% at $3.3 \times 10^{18}\,\mathrm{W} \times \mathrm{cm}^{-2} \times \mu\mathrm{m}^2$ to 80% at $4.2 \times 10^{18}\,\mathrm{W} \times \mathrm{cm}^{-2} \times \mu\mathrm{m}^2$.

Another important feature is the radial expansion velocity of the channel which reached $5 \times 10^8\,\mathrm{cm/s}$ in the first 10 picoseconds and was drastically reduced on a 100-ps timescale. For these experimental conditions, strong MG toroïdal magnetic fields generated by intense currents of electrons with energy in the MeV range were also predicted by 2-D PIC simulations. Such high-energy electrons were actually detected by Malka and co-workers [379] for the same experimental conditions as Fuchs and co-workers. Electrons with energy up to 20 MeV were measured. Two populations of hot electrons with respective temperatures of 1 MeV and 3 MeV were observed on shots at the highest laser power. This feature correlated with a strong increase in laser light transmission and was explained by an increase in laser intensity (and therefore in the associated ponderomotive force) in the plasma channel due to relativistic self-focusing.

In the second phase of the fast igniter scheme, the ponderomotive force associated with the high-intensity laser must push the plasma critical layer close to the high-density core. Perturbation by the ponderomotive force of the critical density surface expansion velocity of a plasma created by the interaction of a short-pulse laser with a massive planar target was first demonstrated by Liu and Umstadter at moderate laser intensity ($\sim 10^{16}\,\mathrm{W/cm^2}$) [387]. In a time-resolved, pump-probe experiment, they studied the Doppler shift of a probe beam reflected onto the critical density surface as a function of the delay between the pump and the probe. They observed that the wavelength of the probe beam was upshifted (redshifted) at the maximum of the pump, and a blueshift was observed before and after the maximum of the pump. This result clearly demonstrates the competition between the ponderomotive and the thermal forces. When the thermal force is higher than the ponderomotive force, the plasma expands in vacuum, and the probe pulse is reflected on a surface moving in the direction of detection. Its wavelength is then blueshifted. Conversely, at maximum pump intensity, the ponderomotive force dominates, and the plasma is pushed inward the target, i.e., in the direction opposite to the detection. Consequently, the wavelength of the probe is redshifted. The recession velocity of the critical density layer was later measured by Kalashnikov and co-workers [396] for laser intensities close to those considered by Wilks and co-workers [134] in their simulations (up to $2 \times 10^{18}\,\mathrm{W/cm^2}$) for pulses with different contrast ratios and for different angles of incidence. Their results were consistent with the model of hole boring proposed by Wilks and co-workers, taking into account conservation of mass and momentum and including the ponderomotive pressure model. Particularly, their experiments showed that for a low contrast pulse ratio ($10^{-3} : 1$), the transition from plasma expansion to hole boring occurs at an intensity below $8 \times 10^{16}\,\mathrm{W/cm^2}$, plasma expansion was still observed at $5 \times 10^{17}\,\mathrm{W/cm^2}$ for a higher contrast ratio ($10^{-7} : 1$). Experimental evidence of hole boring into an overdense plasma was also obtained by Kodama and co-workers [397]. Experiments were performed on preformed plasmas or with massive targets at an intensity of $2 \times 10^{17}\,\mathrm{W/cm^2}$ with the GEKKO-XII laser system [184]. Time-resolved spectra of the backscattered light showed a blueshift at early time related to the plasma expansion, followed by a redshift due to hole boring. The redshift increased with the distance to the target and was only present in the experiments with a preplasma. Confirmation of channel formation was obtained by studying the angular distribution of scattered light. It was observed that backscattered light was collimated when the laser interacts with a preformed plasma, i.e., when a redshift was observed, indicating density channel formation, whereas large spreading of scattered light was measured when using a massive target. Measurements of the hole-boring velocity performed by Kalashnikov and co-workers were further extended to higher laser intensity (up to $10^{19}\,\mathrm{W/cm^2}$) by Zepf and co-workers [398], using 35 TW of the VULCAN laser system [187]. The recession velocity of the critical density surface was deduced from the Doppler shift measured on harmonics of the laser frequency (both odd and even) that are generated by electrons oscillating in the density gradient steepened by the ponderomotive pressure associated with the ultrahigh-intensity laser pulse (see Sect. 11.3). A typical measurement of the Doppler shift in the fourth-harmonic spectrum is shown in Fig. 11.18. The critical density recession velocity found was $v = 0.015 \times c$ at a laser intensity of $10^{19}\,\mathrm{W/cm^2}$. Figure 11.19 shows the recession velocity as a function of the laser intensity, as well as the predictions of the Wilks and co-workers model with (solid line) and without (dashed line) absorption.

Furthermore, interferometric measurements, performed by Tatakiris and co-workers [383], indicated that the hole bored through the plasma persisted long after the laser pulse ($> 1\,\mathrm{ns}$) and that a narrow plasma jet was formed

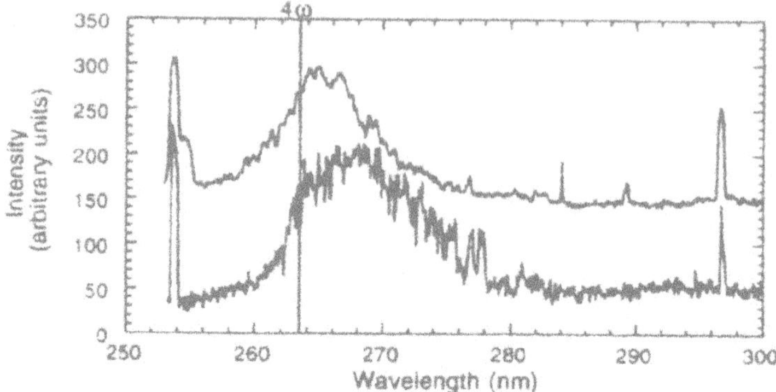

Fig. 11.18. Doppler redshift of the fourth harmonic of the VULCAN Nd:glass laser associated with recession of the plasma critical layer due to ponderomotive pressure [398]

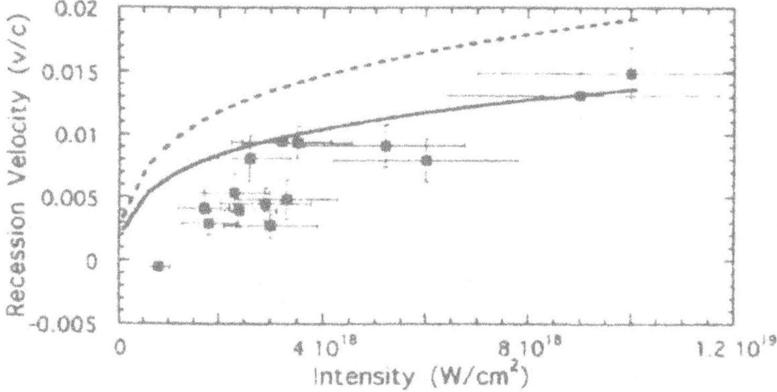

Fig. 11.19. Recession velocity as a function of laser intensity. The solid line is the recession velocity given by the Wilks et al. model, taking into account absorption (60%). The dashed line was calculated without absorption [398]

at the rear side of the target after a few picoseconds, in line with the laser focus. Such behavior was explained by the creation of a beam of fast electrons traveling through the target and collimated inside by the azimuthal magnetic field generated by the fast electrons.

All previous experiments were performed with massive targets and concentrated on the recession of the critical density layer, but Fuchs and co-workers [394,399] recently reported an experimental study of the propagation of the P102 laser [185] (focused at intensity $I = 2\times10^{19}$ W/cm^2) through long (1.5 μm$< l_c <$ 4 μm, where l_c is the distance at which $n_e \geq n_c$) preformed overcritical ($45 < n_e n_c < 100$) plasmas. They measured the laser energy transmitted through the plasma. They observed that up to 10% of the incident energy was transmitted over a distance $l_c \sim 2$ μm. This value was consistent with the hole-boring model proposed by Wilks and co-workers [134].

This experiment demonstrated that a channel can be created in the overdense plasma on a few micron scale length, allowing for the propagation of a fraction of the laser energy through the plasma. One should note that this result is of particular interest for the fast igniter scheme since the laser must penetrate a 70-μm long precompressed plasma with density $n_c < n_e < 10 \times n_c$ before reaching the compressed core [374].

Other observations of light transmitted through overdense plasmas created by the interaction of an ultrahigh laser pulse with a thin foil were also reported [393–395]. In the experiment by Giulietti and co-workers [393], the transmission of the 30-fs, 30TW Ti:sapphire laser system of the Laboratoire d'Optique Appliquée (LOA) [402] through a 0.1-μm plastic foil dramatically increased with the laser intensity and reached about 80% at a laser intensity of 3×10^{18} W/cm^2.

Teychenné and co-workers [403] explained Giulietti and co-workers results by the propagation of the laser light wave in an extraordinary mode through an overdense plasma magnetized by a static field induced by ultrafast ionization of the target material [404]. On the other hand, Fuchs and co-workers [394,395], who performed transmission measurements of the P102 laser system [185] through plastic and aluminum foils of varying thicknesses (from 0.05 to 2 μm), showed that the foil transparency could result from rapid heating of the solid by hot electrons and to the subsequent plasma expansion and density decrease which would allow the laser pulse to propagate by relativistic self-induced transparency through the thin layer. Detection of fast electrons (in the 0.5–3 keV range) on the rear side of the target showed that the number of hot electrons increased with light transmission through the foil [379], which tends to corroborate the interpretation by Fuchs and co-workers.

In the third and final phase of the fast igniter scheme, the fuel is ignited by suprathermal electrons. It is then very important to measure the fraction of laser energy transferred to hot electrons and to study the penetration depth of the electron jet inside a dense material. Numerous experimental studies have been devoted to the subject in the past few years using the well-established technique of K_α-line spectroscopy (see, for example, the review article by Gibbon and Förster and references therein [405]) or bremsstrahlung emission of γ-rays [406–408]. Acceleration of electrons in the MeV range by the ponderomotive force associated with an ultrahigh-intensity laser ($I > 10^{19}$ W/cm^2) was first observed by Malka and Miquel [376] by a direct measurement of the electrons ejected along the laser propagation axis in the forward direction. Using a magnetic spectrometer, they analyzed the energy of electrons transmitted through a thick (~ 30 μm) massive plastic target irradiated by the P102 laser system [185]. The energy distribution of the hot electron tail measured by this method is well fitted by a Boltzmann distribution $f(W) \sim \exp(-W/kT_\mathrm{h})$, where T_h was equal to the energy of the electron oscillating in the laser pulse: $\varepsilon_\mathrm{p} = (\gamma_t - 1) \times m_0c^2$ where

$\gamma_t = (1 + I[\mathrm{W/cm^2}] \times (\lambda[\mu\mathrm{m}])^2/1.37 \times 10^{18})^{1/2}$. The fraction of laser energy transferred to hot electrons (with energy between 0.4 and 3 MeV) was estimated at about 1%, assuming an isotropic distribution.

On the other hand, experimental results obtained by Beg and co-workers [386] with the VULCAN laser system [187] for the same intensity range as Malka and Miquel but for a contrast ratio two decades lower, show a dependence of the hot electron temperature on the laser intensity of the form $T_\mathrm{h} \sim (I\lambda^2)^{1/3}$. This scaling law was deduced from two different diagnostics, K_α-line and γ-ray bremsstrahlung emissions, and was found consistent with a resonance absorption process [409] in the plasma created by the laser pedestal. Furthermore, a compilation by Gibbon and Förster [405] of hot electron temperature measurements, performed by different groups around the world, shows a scaling law with $I\lambda^2$ of the form $T_\mathrm{h} \sim (I\lambda^2)^{1/3-1/2}$ in agreement with PIC simulations. Experiments on fast electron transport have also been recently carried out [380,381] with the petawatt laser of the Lawrence Livermore National Laboratory (LLNL) at intensities up to 10^{20} W/cm^2 [180]. The conversion efficiency, as well as the mean electron energy, and the electron cone-angle were measured by detecting K_α-line X-ray emission for various target materials. Mean electron energy in the 120 –640 keV range was measured, and conversion efficiency was between 20 and 30%, depending on the target material and the laser intensity. The mean electron energy increased with the atomic number of the target. The electron cone angle was measured by imaging the X-ray source onto a charged-coupled-device (CCD) camera set up behind the target, using a penumbral imaging technique [410]. The size of the emitting zone, measured for different target thicknesses, demonstrated the existence of a high-energy (>200 keV) electron beam with a 30° half-cone angle. Such a beaming effect of high-energy electrons was also very recently inferred by Norreys and co-workers [408] from γ-ray measurements. They observed a highly directional and highly energetic (up to 10 MeV) source of γ-rays directly behind the target (see Fig. 11.20).

The γ-rays were produced by bremsstrahlung emission from highly energetic electrons. The emission cone angle roughly corresponded to the bremsstrahlung radiation of electrons with 1–2 MeV energy, undergoing multiple scattering in the dense cold target material. High-energy ions (in the MeV range) associated with the fast-electron-driven plasma expansion were also observed [385,386]. Experiments performed by Fews and co-workers [385] showed that at a laser intensity of 2×10^{18} W/cm^2, 10% of the laser energy can be transferred to ions emitted within a 20° half-cone angle with energies > 100 keV/nucleon and mean energy of 1.3 MeV. Later measurements by Beg and co-workers [386], performed for laser intensities between 10^{17} and 10^{19} W/cm^2, demonstrated that the maximum ion energy associated with fast electrons (fast ions are accelerated by the electrostatic field generated by suprathermal electrons) scaled with laser intensity as $E_\mathrm{max} \sim (I\lambda^2)^{1/3}$. Finally, hot electron transport in a compressed plasma has also been stud-

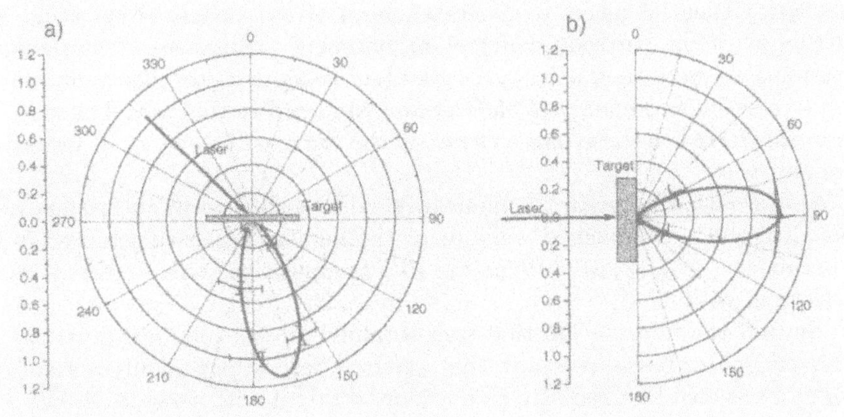

Fig. 11.20. Angular distribution of >10-MeV γ-rays in **(a)** the horizontal and **(b)** the vertical planes [408]

ied. An experiment by Hall and co-workers [384] demonstrated that the fast electron range in a laser-shock-compressed plastic target was approximately twice as large as the range in solid density plastic.

11.8 Pulse Generator of Neutrons

In experimental work relevant to the fast igniter concept, high-energy (2.5 MeV) neutrons produced by deuterated targets irradiated with laser pulses focused at ultrarelativistic intensity (up to 10^{20} W/cm^2) were observed. In this context, neutron emission is used as a diagnostic of the transport of fast ions generated by the interaction of a laser with the target. Neutron energy spectra [380, 411] as well as studies of angular distribution and variation of neutron yield with the laser intensity [412] were reported. The first observation of neutrons generated by the interaction of a high-intensity, short-pulse laser with deuterated targets was made by Norreys and co-workers [411]. Using the VULCAN Nd:glass laser [187] focused at intensities of 10^{19} W/cm^2 onto flat D_2 or C_8D_8 targets, they measured yields of up to 7×10^7 neutrons/sr. Neutron spectroscopy showed that the energy peaked around 2.45 MeV. Measured yields and energy indicate that neutrons were produced by the d(d,n)-^{3}He reaction. Further investigations carried out by Disdier and co-workers [412], using the P102 Nd:glass laser [185] focused at an intensity around 10^{19} W/cm^2 onto deuterated polyethylene planar targets, showed that the neutron yield was strongly dependent on laser intensity.

It was also shown that the emission was anisotropic, with a maximum along the laser propagation axis in the forward direction. This feature suggests a beam–target-type interaction, i.e., a deuteron beam interacts with the thick CD_2 solid target to generate a nonisotropic source of neutrons. The anisotropy of the emission depends on ion energy. Measurements of angular distributions were then used by Disdier and co-workers to infer deuteron energy. An estimated energy of 550 keV was obtained in this way. The laser energy transferred to these ions was nearly the same as the energy in suprathermal electrons.

A correlation between the neutron yield and the energy in fast electrons was also clearly established. Finally, at the highest intensity achieved in the experiment ($\sim 3.5 \times 10^{19}\,\mathrm{W/cm^2}$), $10^7/4\pi$ neutrons were detected in the forward direction.

Beyond the information that was obtained on fast ion transport, the experiments described above show that a pulsed, fast-neutron source can be efficiently produced by irradiating a deuterated target with short, high-intensity laser pulses. Such a neutron source is of particular interest for applications in materials science and neutron radiography [413]. It is one of the reasons that motivated the experiments performed by Ditmire and co-workers [414]. They produced a tabletop source of neutrons using a D_2 cluster jet irradiated by a 35-fs, 120-mJ 10-Hz Ti:sapphire laser focused at $2 \times 10^{16}\,\mathrm{W/cm^2}$. Neutrons were created by the d(d,n)-^{3}He reaction with a total yield of 10^4. The deuteron energy was estimated at 2.5 keV. Such ion energy and the subsequent neutron emission result from the very specific nature of the interaction between a short-pulse, high-intensity laser and large atomic clusters (more than 1000 atoms per cluster). Ionization of clusters results in large ($\sim 80\%$) and rapid ($< 1\,\mathrm{ps}$) collisional absorption of the laser energy by electrons that are heated to many keV [415]. The escape of these hot electrons creates a strong radial electric field that accelerates the ions to very high energy (up to 1 MeV) [416, 417].

Finally, small neutron yields ($\sim 10^2$) were also measured when a 200-mJ 160-fs Ti:sapphire laser was focused at relativistic intensity ($I \sim 10^{18}\,\mathrm{W/cm^2}$) on a near-critical ($n_\mathrm{e} \sim 0.5 \times n_\mathrm{c}$) preformed plasma created by the interaction of a prepulse with a deuterated plastic target [418]. Neutron emission was attributed to fast ions produced by Coulomb explosion of the density channel created by relativistic self-focusing of the high-intensity pulse in the preformed plasma. Note that the space-charge field, resulting from charge separation by the strong ponderomotive force associated with a self-focused laser pulse, can accelerate ions to several MeV, as recently observed by Krushelnick and co-workers [419].

References

1. N. Bloembergen: *Nonlinear Optics* (Addison-Wesley, Redwood City, California 1992)
2. S.A. Akhmanov, V.A. Vysloukh, A.S Chirkin: *Optics of Femtosecond Laser Pulses* (American Institute of Physics, New York 1992)
3. A.P. Sukhorukov: *Nonlinear Wave Interactions in Optics and Radiophysics* (Nauka, Moscow 1988) (in Russian)
4. D. Strickland, G. Mourou: Opt. Commun. **56**, 219 (1985)
5. P. Maine, D. Strickland, P. Bado, M. Pessot, G. Mourou: IEEE J. Quantum Electron. QE-**24**, 398 (1988)
6. B. Luther-Davis, E.G. Gamaliy, Wang Yanji, A.B. Rode, V.T. Tikhonchuk: Kvantovaya Elektronika **19**, 317 (1992) [Sov. J. Quantum. Electron. **22**(4), 289 (1992)]
7. A.B. Borisov, A.V. Borovskiy, O.B. Shiryaev, V.V. Korobkin, A.M. Prokhorov, J.C. Solem, T.S. Luk, K. Boyer, C.K. Rhodes: Phys. Rev. A **45**, 5830 (1992)
8. A.B. Borisov, A.V. Borovskiy, V.V. Korobkin, A.M. Prokhorov, O.B. Shiryaev, X.M. Shi, T.S. Luk, A. McPherson, J.C. Solem, K. Boyer, C.K. Rhodes: Phys. Rev. Lett. **68**, 2309 (1992)
9. N.L. Tsintstadze, D.D. Tskhakaya: *Relativistic Nonlinear Effects in Plasmas* (Metsniereba, Tbilisi 1989) (in Russian)
10. W.L. Kruer: *The Physics of Laser Plasma Interactions* (Addison-Wesley, Redwood City, California 1988)
11. N.I. Koroteev, I.L. Shumay: *Physics of Powerful Laser Radiation* (Nauka, Moscow 1991) (in Russian)
12. A.V. Borovsky, A.L. Galkin: *Laser Physics* (IzdAT, Moscow 1996) (in Russian)
13. G.A. Askar'yan: Zh. Eksp. Teor. Fiz. **42**, 1672 (1962) [Sov. Phys. JETP **15**(6), 1161 (1962)]
14. M. Hercher: J. Opt. Soc. Am. **54**, 563 (1964)
15. N.F. Pilipetsky, A.F. Rustamov: Pis'ma v Zh. Eksp. Teor. Fiz. **2**, 88 (1965) [JETP Lett. **2**(2), 55 (1965)]
16. V.I. Talanov: Pis'ma v Zh. Eksp. Teor. Fiz. **2**, 218 (1965) [JETP Lett. **2**(5), 138 (1965)]
17. A.G. Litvak, V.I. Talanov: Izvestiya Vuzov Radiofizika **10**, 539 (1967) [Sov. Radiophysics **10**(4), 296 (1967)]
18. R.Y. Chiao, E. Garmire, C.H. Townes: Phys. Rev. Lett. **13**, 479 (1964)
19. H.A. Haus: Appl. Phys. Lett. **8**, 128 (1966)
20. Z.K. Yankauskas: Izvestiya Vuzov Radiofizika **9**, 412 (1966) [Sov. Radiophysics **9**(2), 261 (1966)]
21. V.E. Zakharov, A.B. Shabat: Zh. Eksp. Teor. Fiz. **61**, 118 (1971) [Sov. Phys. JETP **34**(1), 62 (1972)]

22. V.E. Zakharov, S.V. Manakov: Zh. Eksp. Teor. Fiz. **71**, 203 (1976) [Sov. Phys. JETP **44**(1), 106 (1976)]
23. V.I. Bespalov, V.I. Talanov: Pis'ma v Zh. Eksp. Teor. Fiz. **3**, 471 (1966) [JETP Lett. **3**(12), 307 (1966)]
24. V.E. Zakharov: Zh. Eksp. Teor. Fiz. **53**, 1735 (1967) [Sov. Phys. JETP **26**(5), 994 (1968)]
25. V.E. Zakharov, A.M. Rubenchik: Zh. Eksp. Teor. Fiz. **65**, 999 (1973) [Sov. Phys. JETP **38**(3), 494 (1974)]
26. N.G. Vakhitov, A.A. Kolokolov: Izvestiya Vuzov. Radiofizika **16**, 1020 (1973) [Sov. Radiophysics **16**(7), 783 (1973)]
27. V.I. Talanov: Izvestiya Vuzov. Radiofizika **9**, 410 (1966) [Sov. Radiophysics **9**(2), 260 (1966)]
28. O.I. Kudryashov: Sibirsky Matematichesky Zhurnal **16**, 866 (1975) (in Russian)
29. V.E. Zakharov, V.V. Sobolev, V.S. Synakh: Pis'ma v Zh. Eksp. Teor. Fiz. **14**, 564 (1971) [JETP Lett. **14**(10), 390 (1971)]
30. V.E. Zakharov, V.S. Synakh: Zh. Eksp. Teor. Fiz. **68**, 940 (1975) [Sov. Phys. JETP **41**(3), 465 (1976)]
31. V.E. Zakharov, V.V. Sobolev, V.S. Synakh: Zh. Eksp. Teor. Fiz. **60**, 136 (1971) [Sov. Phys. JETP **33**(1), 77 (1971)]
32. L.D. Landau, E.M. Lifshitz: *The Classical Field Theory* (Pergamon Press, Oxford 1962)
33. L.S. Brown, T.W.B. Kibble: Phys. Rev. **133**, 705 (1964)
34. S. Sarachik, G.T. Shappert: Phys. Rev. D **1**(10), 2738 (1970)
35. C. I. Castillo–Herrera, T.W. Johnston: IEEE Trans. Plasma Sci. **21**, 125 (1993)
36. J. Krüger, M. Bovyn: J. Phys. A **9**, 1841 (1976)
37. J.N. Bardsley, B.M. Penetrante, M.H. Mittleman: Phys. Rev. A **40**(7), 3823 (1989)
38. S.P. Goreslavskii, N.B. Narozhniy, V.P. Yakovlev: Laser Phys. **1**(6), 670 (1991)
39. A.I. Akhiezer, R.V. Polovin: Zh. Eksp. Teor. Fiz. **30**, 915 (1956) [Sov. Phys. JETP **3**(5), 696 (1956)]
40. C. Max, F.W. Perkins: Phys. Rev. Lett. **27**, 1342 (1971)
41. A.C.L. Chian, P.C. Clemmow: J. Plasma Phys. **14**, 505 (1975)
42. P.K. Kaw, A. Sen, E.J. Valeo: Physics D **9**, 96 (1983)
43. A.V. Borovsky, A.L. Galkin, O.B. Shiryaev: Kratkie Sobsheniya po Fizike **5**, 33 (1998) [Bulletin of the Lebedev Physics Institute **5**, 28 (1998)]
44. A.V. Borovsky, A.L. Galkin, V.V. Korobkin, O.B. Shiryaev: Phys. Rev. E **59**, 2253 (1998)
45. V.A. Kozlov, A.G. Litvak, E.V. Suvorov: Zh. Eksp. Teor. Fiz. **76**, 148 (1979) [Sov. Phys. JETP **49**(1), 75 (1979)]
46. P.K. Kaw, A. Sen, T. Katsouleas: Phys. Rev. Lett. **68**, 3172 (1992)
47. R.N. Sudan, Y.S. Dimant, O.B. Shiryaev: Phys. Plasmas **4**(5), 1489 (1997)
48. R.J. Noble: Phys. Rev. A **32**, 460 (1985)
49. A. Magneville: J. Plasma Phys. **44**, 231 (1990)
50. H. Hora: *Physics of Laser Driven Plasmas* (Wiley, New York 1981)
51. V.E. Zakharov: Zh. Eksp. Teor. Fiz. **60**, 1714 (1971) [Sov. Phys. JETP **33**(5), 927 (1971)]
52. S. Jorna: Phys. Fluids **17**, 765 (1973)
53. J.F. Drake, P.K. Kaw, Y.C. Lee, G. Schmidt, C.S. Liu, M.N. Rosenbluth: Phys. Fluids **17**, 78 (1974)
54. T.M. Antonsen, P. Mora: Phys. Fluids B **5**(5), 1440 (1993)
55. C. Joshi, T. Tajima, J.M. Dawson, H.A. Baldis, N.A. Ebrahim: Phys. Rev. Lett. **47**, 1285 (1981)

56. C.J. McKinstrie, A. Simon, E.A. Williams: Phys. Fluids **27**, 2738 (1984)
57. V.K. Tripathi, C.S. Liu: Phys. Fluids B **3**, 468 (1990)
58. P. Sprangle, E. Esarey: Phys. Rev. Lett. **67**(15), 2021 (1991)
59. W.B. Mori, C.D. Decker, D.E. Hinkel, T. Katsouleas: Phys. Rev. Lett. **72**, 1482 (1994)
60. A.M. Kalmykov, N.Ya. Kotsarenko: Izvestiya Vuzov Radiofizika **19**, 1481 (1976) [Sov. Radiophysics **19**(10), 1038 (1976)]
61. V.I. Kirsanov, A.S. Sakharov: Fizika Plazmy **21**, 623 (1995) [Plasma Phys. Rep. **21**(7), 587 (1995)]
62. A.S. Sakharov, V.I. Kirsanov: Fizika Plazmy **21**, 632 (1995) [Plasma Phys. Rep. **21**(7), 596 (1995)]
63. A.V. Borovsky, V.V. Korobkin, O.B. Shiryaev, A.L. Galkin: Zh. Eksp. Teor. Fiz. **113**, 2034 (1998) [Sov. Phys. JETP **86**, 6 (1998)]
64. B. Quesnel, P. Mora, J.C. Adam, S. Guerin, A. Heron, G. Laval: Phys. Rev. Lett. **78**, 2132 (1997)
65. B. Quesnel, P. Mora, J.C. Adam, A. Heron, G. Laval: Phys. Plasmas **4**, 3358 (1997)
66. A.S. Sakharov, V.I. Kirsanov: Phys. Plasmas **4**(9), 3382 (1997)
67. A.V. Borovsky, A.L. Galkin, V.V. Korobkin, O.B. Shiryaev: Kvantovaya Elecron. **26**, 49 (1999) [Quantum Electronics **29**, 49 (1999)]
68. A.V. Borovsky, A.L. Galkin, V.V. Korobkin, O.B. Shiryaev: Zh. Eksp. Teor. Fiz. **116**(6), 1947 (1999) [Sov. Phys. JETP **88**(6), 1055 (1999)]
69. K. Krushelnick, A. Ting, C.I. Moore, H.R. Burris, E. Esarey, P. Sprangle, M. Baine: Phys. Rev. Lett. **78**, 4047 (1997)
70. M. Watanabe, K. Hata, T. Adachi, R. Nodomi, S. Watanabe: Opt. Lett. **15**, 845 (1990)
71. A.J. Taylor, C.R. Tallman, J.P. Roberts, C.S. Lester, T.R. Gosnell, P.H.Y. Lee, G.A. Kyrala: Opt. Lett. **15**, 39 (1990)
72. M.D. Perry, E.M. Campbell, J.T. Hunt, G. Heane, A. Szoke, G. Mourou, P. Bado, P. Maine: Ultra-High Brightness Laser Facility. LLNL, CA (1987)
73. G. Mourou, D. Umstadter: Phys. Fluids B **4**, 2315 (1992)
74. N.B. Delone, V.P. Krainov: *Atoms in Strong Light Fields* (Springer-Verlag, Berlin, New York 1985)
75. T.S. Luk, U. Johann, H. Egger, H. Pummer, C.K. Rhodes: Phys. Rev. A **32**, 214 (1985)
76. U. Johann, T. Luk, H. Egger, C.K. Rhodes: Phys. Rev. A **34**, 1084 (1986)
77. C.E. Max, J. Arons, B. Langdon: Phys. Rev. Lett. **33**, 209 (1974)
78. C.E. Max: Phys. Fluids **19**, 74 (1976)
79. G. Schmidt , W. Horton: Comments Plasma Phys. Controlled Fusion **9**, 85 (1985)
80. A.Zh. Muradyan: Zh. Eksp. Teor. Fiz. **92**, 1978 (1987) [Sov. Phys. JETP **65**(6), 1111 (1987)]
81. Sun Guo-Zheng, E. Ott, Y.C. Lee, P. Guzdar: Phys. Fluids **30**, 526 (1987)
82. X.L. Chen, R.N. Sudan: Phys. Rev. Lett. **70**, 2082 (1993)
83. P. Sprangle, E. Esarey, A. Ting, G. Joyce: Appl. Phys. Lett. **53**(22), 2146 (1988)
84. P. Sprangle, E. Esarey, A. Ting: Phys. Rev. Lett. **64**, 2011 (1990)
85. A. Ting, E. Esarey, P. Sprangle: Phys. Fluids B **2**, 1390 (1990)
86. P. Sprangle, A. Zigler, E. Esarey: Appl. Phys. Lett. **58**(4), 346 (1991)
87. P. Sprangle, E. Esarey, J. Krall, G. Joyce: Phys. Rev. Lett. **69**(15), 2200 (1992)
88. P. Sprangle, E. Esarey, J. Krall: Phys. Plasmas **3**(5), 2183 (1996)
89. T. Kurki-Suonio, J. Morrison, T. Tajima: Phys. Rev. A **40**, 3230 (1989)

90. A.V. Gaponov, M.A. Miller: Zh. Eksp. Teor. Fiz. **34**, 242 (1958) [Sov. Phys. JETP **7**, 168 (1958)]
91. A.V. Gaponov, M.A. Miller: Zh. Eksp. Teor. Fiz. **34**, 751 (1958) [Sov. Phys. JETP **7**, 515 (1958)]
92. K. Nakajima, D. Fisher, T. Kawakubo, H. Nakanishi, A. Ogata, Y. Kato, Y. Kitagawa, R. Kodama, K. Mima, H. Shiraga, K. Suzuki, K. Yamakawa, T. Zhang, Y. Sakawa, T. Shiji, Y. Nishida, N. Yugami, M. Downer, T. Tajima: Phys. Rev. Lett. **74**, 4428 (1995)
93. W.B. Mori, C. Joshi, J.M. Dawson, D.W. Forslund, J.M. Kindel: Phys. Rev. Lett. **60**, 1298 (1988)
94. A.B. Borisov, A.V. Borovskiy, V.V. Korobkin, A.M. Prokhorov, C.K. Rhodes, O.B. Shiryaev: Phys. Rev. Lett. **65**, 1753 (1990)
95. A.B. Borisov, A.V. Borovskiy, V.V. Korobkin, A.M. Prokhorov, C.K. Rhodes, O.B. Shiryaev: Zh. Eksp. Teor. Fiz. **101**, 1132 (1992) [Sov. Phys. JETP **74**(4), 604 (1992)]
96. A.B. Borisov, A.V. Borovskiy, V.V. Korobkin, A.M. Prokhorov, O.B. Shiryaev, C.K. Rhodes: Laser Phys. **1**(5), 546 (1991)
97. A.B. Borisov, O.B. Shiryaev, A. McPherson, K. Boyer, C.K. Rhodes: Comments Plasma Phys. Controlled Fusion **37**, 564 (1995)
98. A.V. Borovskiy, Ya.M. Zhileikin, Yu.I. Osipik, E.A. Makarova: Laser Phys. **7**, 485 (1997)
99. A.B. Borisov, A.V. Borovskiy, V.V. Korobkin, A.M. Prokhorov, O.B. Shiryaev, J.C. Solem, A. McPherson, K. Boyer, C.K. Rhodes: J. Opt. Soc. Am. B **11**, 1941 (1994)
100. A.V. Borovskiy, Ya.M. Zhileikin, V.V. Korobkin, E.A. Makarova, Yu.I. Osipik, A.M. Prokhorov: Laser Phys. **4**, 1173 (1994)
101. A.V. Borovskiy, Ya.M. Zhileikin, V.V. Korobkin: Kvantovaya Elektronika **25**, 1 (1995) [Quantum Electron. **25**(4), 366 (1995)]
102. A.V. Borovskiy, A.L. Galkin, Ya.M. Zhileikin, E.A. Makarova, Yu.I. Osipik: *Nonlinear Propagation of Ultrashort Powerful Laser Pulses in Matter* (Moscow State University Publishing House, Moscow 1995)
103. A.V. Borovskiy, A.L. Galkin: Zh. Eksp. Teor. Fiz. **104**, 3311 (1993) [Sov. Phys. JETP **77**(4), 562(1993)]
104. S.V. Bulanov, V.I. Kirsanov, A.S. Sakharov: Fizika Plazmy **16**, 935 (1990) [Sov. J. Plasma Phys. **16**(8), 543 (1990)]
105. S.V. Bulanov, A.S. Sakharov: Pis'ma v Zh. Eksp. Teor. Fiz. **54**, 208 (1991) [JETP Lett. **54**(4), 203 (1991)]
106. S.V. Bulanov, I.N. Inovenkov, V.I. Kirsanov, N.M. Naumova, A.S. Sakharov: Phys. Fluids B **4**, 1935 (1992)
107. S.V. Bulanov, F. Pegoraro: Laser Phys. **6**, 1120 (1994)
108. S.V. Bulanov, N.M. Naumova, F. Pegoraro: Phys. Plasmas **1**, 745 (1994)
109. S.V. Bulanov, N.M. Naumova, F. Pegoraro: Fizika Plazmy **20**, 640 (1994) [Plasma Phys. Rep. **20**(7), 574 (1994)]
110. S.V. Bulanov, F.F. Kamenets, F. Pegoraro, A.M. Pukhov: Phys. Lett. A **195**, 84 (1994)
111. S.V. Bulanov, M. Lontano, T.Zh. Esirkepov, F. Pegoraro, A.M. Pukhov: Phys. Rev. Lett. **76**, 3562 (1996)
112. S.V. Bulanov, V.A. Vshivkov, G.I. Dudnikova, N.M. Naumova, F. Pegoraro, I.V. Pogorelsky: Fizika Plazmy **23**(4), 284 (1997) [Plasma Phys. Rep. **23**(4), 259 (1997)]
113. N.E. Andreev, L.M. Gorbunov, V.I. Kirsanov, A.A. Pogosova, R.R. Ramazashvili: Pis'ma v Zh. Eksp. Teor. Fiz. **55**, 550 (1992) [JETP Lett. **55**(10), 571 (1992)]

114. N.E. Andreev, L.M. Gorbunov, V.I. Kirsanov, A.A. Pogosova: Pis'ma v Zh. Eksp. Teor. Fiz. **60**, 694 (1994) [JETP Lett. **60**(10), 713 (1994)]
115. A.V. Borovsky, A.L. Galkin: Zh. Eksp. Teor. Fiz. **106**, 915 (1994) [Sov. Phys. JETP **79**(3), 502 (1994)]
116. A.V. Borovskiy, A.L. Galkin, V.B. Karpov: Laser Phys. **6**(4), 785 (1996)
117. A.V. Borovskiy, A.L. Galkin, V.B. Karpov: Proc. Gen. Phys. Inst. Russ. Acad. Sci. **50**, 83 (1995) (in Russian)
118. A.V. Borovskiy, A.L. Galkin: Zh. Eksp. Teor. Fiz. **109**, 1070 (1996) [Sov. Phys. JETP **82**(3), 576 (1996)]
119. X.L. Chen, R.N. Sudan: Phys. Fluids B **5**, 1336 (1993)
120. A.V. Borovskii, O.B. Shiryaev: Zh. Eksp. Teor. Fiz. **109**, 865 (1996) [Sov. Phys. JETP **83**(3), 475 (1996)]
121. A.V. Borovskii, A.L. Galkin, V.V. Korobkin, O.B. Shiryaev: Kvantovaya Elektronika **24**(10), 929 (1997) [Quantum Electronics **27**(10), 903 (1997)]
122. K. Boyer, T.S. Luk, A. McPherson, X. Shi, J.C. Solem, C.K. Rhodes, A.B. Borisov, A.V. Borovsky, O.B. Shiryaev, V.V. Korobkin: X-Ray Amplifier Energy Deposition Scaling with Channeled Propagation. In: *Proceedings of the International Conference on Lasers,* December 9 - 13, 1991
123. A.B. Borisov, A.V. Borovsky, O.B. Shiryaev, V.B. Karpov, V.V. Korobkin, A.M. Prokhorov, J.C. Solem, A. McPherson, X. Shi, T.S. Luk, K. Boyer, C.K. Rhodes: (1992) Investigation of Relativistic and Charge-Displacement Self - Channeling of Intense Subpicosecond Ultraviolet (248 nm) Radiation in Plasmas. In: *Inst. Phys. Conf. Ser. No 125: Section 4, Paper Presented at Int. Colloquium on X-Ray Lasers.* Schiliersee, Germany 1992
124. A.B. Borisov, A.V. Borovskiy, O.B. Shiryaev, D.A. Tate, B.E. Bouma, X. Shi, A. McPherson, T.S. Luk, C.K. Rhodes: Appl. Opt. **31**, 3433 (1992)
125. A.V. Borovskiy, V.V. Korobkin, A.M. Prokhorov: Zh. Eksp. Teor. Fiz. **106**, 148 (1994) [Sov. Phys. JETP **79**(1), 81 (1994)]
126. V.B. Gildenburg, A.V. Kim, V.A. Krupnov, V.E. Semenov, A.M. Sergeev, N.N. Zharova: IEEE Trans. Plasma Sci. **21**(1), 34 (1993)
127. R.L. Savage Jr., R.P. Brogle, W.B. Mori, C. Joshi: IEEE Trans. Plasma Sci. **21**(1), 5 (1993)
128. Ya.L. Bogomolov, S.F. Lirin, V.E. Semenov, A.M. Sergeev: Pis'ma v Zh. Eksp. Teor. Fiz. **45**, 532 (1987) [JETP Lett. **45**(11), 680 (1987)]
129. A.V. Borovskiy, A.L. Galkin: Zh. Eksp. Teor. Fiz. **108**, 426 (1995) [Sov. Phys. JETP **81**(2), 230 (1995)]
130. E.G. Gamaliy, R. Dragila: Phys. Rev. A **42**, 929 (1990)
131. T.Zh. Esirkepov, F.F. Kamenets, S.V. Bulanov, N.M. Naumova: Pis'ma v Zh. Eksp. Teor. Fiz. **68**(1), 33 (1998) [JETP Lett. **68**(1), 36 (1998)]
132. J.H. Marburger, R.F. Tooper: Phys. Rev. Lett. **35**, 1001 (1975)
133. C.S. Lai: Phys. Rev. Lett. **36**, 966 (1976)
134. S.C. Wilks, W.L. Kruer, M. Tabak, A.B. Langdon: Phys. Rev. Lett. **69**, 1383 (1992)
135. R.N. Sudan: Phys. Rev. Lett. **70**, 3075 (1993)
136. V.B. Berestetsky, E.M. Lifshits, L.P. Pitaevsky: *Quantum Electrodynamics* (Pergamon Press, Oxford; New York 1982)
137. M.V. Fedorov: *Atomic and Free Electrons in a Strong Light Field* (World Scientific, Singapore, River Edge, NJ 1997)
138. A.V. Borovskiy, V.V. Korobkin, A.M. Prokhorov: Laser Phys. **3**, 713 (1993)
139. R.L. Seliger, G.B. Witham: Proc. R. Soc. London Ser. A **305**, 1 (1968)
140. L.D. Landau, E.M. Lifshitz: *Mechanics* (Pergamon Press, Oxford 1960)
141. H.H. Kuehl, C.Y. Zhang, T. Katsouleas: Phys. Rev. E **47**, 1249 (1993)
142. H.H. Kuehl, C.Y. Zhang: Phys. Rev. E **48**, 1316 (1993)

143. J.H. Wilkinson, S. Reinsch: *Linear Algebra* (Springer-Verlag, Berlin 1971)
144. V.N. Tsytovich: *Theory of Turbulent Plasma* (Consultants Bureau, NY 1977)
145. V.P. Silin: *Parametric Effect of High Power Radiation on Plasma* (Nauka, Moscow 1973) (in Russian)
146. Yu.L. Klimontovich: *Statistical Physics* (Harwood Academic, NY 1986)
147. L.M. Gorbunov. V.I. Kirsanov: Proc. Lebedev Phys. Inst. Russ. Acad. Sci. **219**, 3 (1992) (in Russian)
148. A.J. Alcock, C. Demichelis, V.V. Korobkin, M.C. Richardson: Phys. Lett. A **29**, 475 (1969)
149. N.I. Il'ichev, V.V. Korobkin, V.A. Korshunov, et al.: Zh. Eksp. Teor. Fiz. **15**, 191 (1972)
150. V.V. Korobkin, A.M. Prokhorov, R.V. Serov, et al.: Phys. Lett. A **47**, 381 (1974)
151. A.V. Borovskiy, Ya.M. Zhileykin, V.V. Korobkin, et al.: Laser Phys. **5**, 868 (1995)
152. T.S. Luk, A. McPherson, G. Gibson, K. Boyer, C.K. Rhodes: Opt. Lett. **14**, 1113 (1989)
153. S. Augst, D. Strickland, D.D. Meyerhofer, S.L. Chin, J. Eberly: Phys. Rev. Lett. **63**, 2212 (1989)
154. G. Gibson, T.S. Luk, C.K. Rhodes: Phys. Rev. A **41**, 5049 (1990)
155. P. Monot, T. Auguste, L.A. Lompré, et. al.: J. Opt. Soc. Am. B , 1579 (1992)
156. R.W. Falcone: In: *X - Ray Lasers, Third International Colloquium on X-ray Lasers,* Schliersee, Germany, 1992 ed. by E. Fill (Institute of Physics, Bristol, Philadelphia 1992) p. 213
157. J.R. Marques, F. Amiranoff, A. Dyson, et al.: Phys. Fluids B **5**, 597 (1993)
158. V.P. Kandidov, O.G. Kosareva, S.A. Shlyonov: Kvantovaya Electronika **21**, 971 (1994) [Quantum Electron.]
159. M.V. Ammosov, N.B. Delone, V.P. Krainov: J. Exp. Teor. Fiz. **91**, 2008 (1986) [Sov. Phys. JETP]
160. L.A. Vainshtein, I.I. Sobel'man, E.A. Yukov: *Excitation of Atoms and Broadening of Spectral Lines* (Springer-Verlag, Berlin, New York 1981)
161. A.V. Borovsky, A.L. Galkin, V.G. Priymak: Zhurnal Vychislitel'noy Matematiki i Matematicheskoy Fiziki **35**, 565 (1995) [Comp. Math. Math. Phys. **35**(4), 447 (1995)]
162. A.V. Borovsky, A.L. Galkin, V.G. Priymak, E.V. Chizhonkov: (1990) Zhurnal Vychislitel'noy Matematiki i Matematicheskoy Fiziki **30**, 1381 (1990) [Comp. Math. Math. Phys. (1995)]
163. A.B. Borisov, A.V. Borovsky, V.B. Karpov, et al: In: *X - Ray Lasers, Third International Colloquium on X-ray Lasers,* Schliersee, Germany, 1992, ed. by E. Fill (Institute of Physics, Bristol; Philadelphia 1992)
164. A.V. Borovsky, S.A. Zapryagaev, O.I. Zatsarinny, N.L. Manakov: *Plasmas of Multiple Charge Ions* (Khimiya, Saint-Petersburg 1995) (in Russian)
165. T. Auguste, P. Monot, L.A. Lompré, G. Mainfray, and C. Manus: Opt. Commun. **89**, 145 (1992)
166. S.C. Rae: Opt. Commun. **104**, 330 (1994)
167. E.E. Fill: J. Opt. Soc. Am. **B 11**, 2241 (1994)
168. P.R. Bolton, A.B. Bullock, C.D. Decker, M.D. Feit, A.J.P. Megofna, P.E. Young, D.N. Fittinghoff: J. Opt. Soc. Am. B **13**, 336 (1996)
169. A.J. Mackinnon, M. Borghesi, A. Iwase, M.W. Jones, G.J. Pert, S. Rae, K. Burnett, O. Willi: Phys. Rev. Lett. **76**, 1473 (1996)
170. S.P. Nikitin, Y. Li, T. M. Antonsen, H.M. Milchberg: Opt. Commun. **157**, 139 (1998)

171. P. Chessa, E. De Wispelaere, F. Dorchies, V. Malka, J.R. Marquès, G. Harmoniaux, P. Mora, F. Amiranoff: Phys. Rev. Lett. **82**, 552 (1999)
172. E. Yablonovitch: Phys. Rev. Lett. **32**, 1101 (1974)
173. W.M. Wood, G. Focht, M.C. Downer: Opt. Lett. **13**, 984 (1988)
174. W.M. Wood, C.W. Siders, M.C. Downer: Phys. Rev. Lett **67**, 3523 (1991)
175. S.P. Le Blanc, R. Sauerbrey, S.C. Rae, K. Burnett: J. Opt. Soc. Am. **B 10**, 1801 (1993)
176. M. Ciarrocca, J.P. Marangos, D.D. Burgess, M.H.R. Hutchinson, R.A. Smith, S.C. Rae, K. Burnett: Opt. Commun. **110**, 425 (1994)
177. A. Sullivan, H. Hamster, S.P. Gordon, R.W. Falcone, H. Nathel: Opt. Lett. **19**, 1544 (1994)
178. S.P. Le Blanc, and R. Sauerbrey: J. Opt. Soc. Am. **B 13**, 72 (1996)
179. K. Yamakawa, M. Aoyama, S. Matsuoka, Y. Akahane, H. Takuma: Opt. Lett. **23**, 18 (1998)
180. M. D. Perry, D. Pennington, B.C. Stuart, G. Tiebohl, J.A. Britten, C. Brown, S. Herman, B. Golick, M. Kartz, J. Miller, H.T. Powell, M. Vertino, V. Yanovsky: Opt. Lett. **24**, 3 (1999)
181. A. Ting, K. Krushelnick, C.I. Moore, H.R. Burris, E. Esarey, J. Krall, P. Sprangle: Phys. Rev. Lett.**77**, 5377 (1996)
182. Y.H. Chuang, D.D. Meyerhofer, S. Augst, H. Chen, J. Peatross, S. Uchida: J. Opt. Soc. Am. B **8**, 1226 (1991)
183. D. Umstadter, S.Y. Chen, A. Maksimchuk, G. Mourou, R. Wagner: Science **273**, 472 (1996)
184. Y. Kitagawa, R. Kodama, K. Takahashi, M. Mori, M. Iwata, S. Tuji, K. Suzuki, K. Sawai, K. Hamada, K. Tanaka, H. Fujita, T. Kanabe, H. Takabe, H. Habara, Y. Kato, K. Mima: Fusion Eng. Design **44**, 261 (1999)
185. N. Blanchot, C. Rouyer, S. Sauteret, A. Migus: Opt. Lett. **20**, 395 (1995)
186. J.P. Zou, S. Savalle, L. Martin, P. Moreau, A.M. Sautivet: Etude de l'amplification et du spectre dans la chaîne multi-verres. LULI Internal Report 1997 (Ecole Polytechnique Palaiseau, France, 1998, NTIS: PB98 - 152515) (in French)
187. C.N. Danson, L.J. Barzanti, Z. Chang, A.E. Damerell, C.B. Edwards, S. Hancock, M.H.R. Hutchinson, M.H. Key, S. Luan, R.R. Mahadeo, I.P. Mercer, P. Norreys, D.A. Pepler, D.A. Rodkiss, I.N. Ross, M.A. Smith, R.A. Smith, P. Taday, W.T. Toner, K.W.M. Wigmore, T.B. Winstone, R.W.W. Wyatt, F. Zhou: Opt. Commun. **103**, 392 (1993)
188. M.P. Kalashnikov, G. Sommerer, P.V. Nickles, W. Sandner: Opt. Commun. **133**, 216 (1997)
189. S. Sartania, Z. Cheng, M. Lenzner, G. Tempea, Ch. Spielmann, F. Krausz, K. Ferencz: Opt. Lett. **22**, 1562 (1997)
190. C. Le-Blanc, E. Baubeau, F. Salin, J.A. Squier, C.P.J. Barty, C. Spielmann: IEEE J. Quantum Electron. **4**, 407 (1998)
191. C.G. Durfee, S. Backus, M.M. Murnane, H.C. Kapteyn: IEEE J. Quantum Electron. **4**, 395 (1998)
192. C.G. Durfee III, H.M. Milchberg: Phys. Rev. Lett. **71**, 2409 (1993)
193. C.G. Durfee III, J. Lynch, H.M. Milchberg: Phys. Rev. E **51**, 2368 (1995)
194. T.R. Clark, H.M. Milchberg: Phys. Rev. Lett. **78**, 2373 (1997)
195. S.P Nikitin, T.M. Antonsen, T.R Clark, Yuelin Li, H.M. Milchberg: Opt. Lett. **22**, 1797 (1997)
196. S.P. Nikitin, I. Alexeev, J. Fan, H.M. Milchberg: Phys. Rev. E **59**, R3839 (1999)
197. J. Fan, T.R. Clark, H.M. Milchberg: Appl. Phys. Lett. **73**, 3064 (1998)
198. P. Volfbeyn, E. Esarey, W.P. Leemans: Phys. Plasmas **6**, 2269 (1999)

199. A. Zigler, Y. Ehrlich, C. Cohen, J. Krall, P. Sprangle: J. Opt. Soc. Am. B **13**, 68 (1996)
200. Y. Ehrlich, C. Cohen, A. Zigler, J. Krall, P. Sprangle, E. Esarey: Phys. Rev. Lett. **77**, 4186 (1996)
201. Y. Ehrlich, C. Cohen, D. Kaganovich, A. Zigler, R.F. Hubbard, P. Sprangle, E. Esarey: J. Opt. Soc. Am. B **15**, 2416 (1998)
202. S. Jackel, R. Burris, J. Grun, A. Ting, C. Manka, K. Evans, J. Kosakowskii: Opt. Lett. **20**, 1086 (1995)
203. M. Borghesi, A.J. Mackinnon, R. Gaillard, O. Willi, A.A. Offenberger: Phys. Rev. E **57**, R4899 (1998)
204. F. Dorchies, J.R. Marquès, B. Cros, G. Matthieussent, C. Courtois, T. Vélikoroussov, P. Audebert, J.P. Geindre, S. Rebibo, G. Hamoniaux, F. Amiranoff: Phys. Rev. Lett. **82**, 4655 (1999)
205. J.H. Mcleod: J. Opt. Soc. Am. **44**, 592 (1954)
206. N.H. Burnett, P.B. Corkum: J. Opt. Soc. Am. B **6**, 1195 (1989)
207. N.H. Burnett, G.D. Enright: IEEE J. Quantum Electron. **26**, 1797 (1990)
208. P. Amendt, D.C. Eder, S.C. Wilks: Phys. Rev. Lett. **66**, 2589 (1991)
209. D.C. Eder, P. Amendt, S.C. Wilks: Phys. Rev. A **45**, 6761 (1992)
210. P. Amendt, D.C. Eder, R.A. London, M.D. Rosen: Phys. Rev. A **47**, 1572 (1993)
211. D.C. Eder, P. Amendt, L.B. DaSilva, R.A. London, B.J. MacGowan, D.L. Matthews, B.M. Penetrante, M.D. Rosen, S.C. Wilks, T.D. Donnelly, R.W. Falcone, G.L. Strobel: Phys. Plasmas **1**, 1744 (1994)
212. *High Resolution Laser Photoionization and Photoelectron Studies.* In: Wiley Series in Ion Chemistry and Physics, ed. by I. Powis, T. Baer, C.Y. Ng (Wiley, Chichester, England, New York 1995)
213. M. Gentili, C. Giovannella, S. Selci: Nanolithography: A Borderland between STM, EB, IB, and X-Ray Lithographies. NATO ASI Series E: Appl. Sci. 264 (1994)
214. S. Augst, D.D. Meyerhofer, D. Strickland, S.L. Chin: J. Opt. Soc. Am. B **8**, 858 (1991)
215. T. Auguste, P. Monot, L.A. Lompré, C. Manus, G. Mainfray: J. Phys. B: At. Mol. Opt. Phys. **25**, 4181 (1992)
216. A.A. Offenberger, W. Blyth, A.E. Dangor, A. Djaoui, M.H. Key, Z. Najmudin, J.S. Wark: Phys. Rev. Lett. **71**, 3983 (1993)
217. T.E. Glover, T.D. Donnelly, E.A. Lipman, A. Sullivan, R.W. Falcone: Phys. Rev. Lett. **73**, 78 (1994)
218. W.J. Blyth, S.G. Preston, A.A. Offenberger, M.H. Key, J.S. Wark, Z. Najmudin, A. Modena, A. Djaoui, A.E. Dangor: Phys. Rev. Lett. **74**, 554 (1995)
219. A.A. Offenberger, W. Blyth, A.E. Dangor, A. Djaoui, M.H. Key, A. Modena, Z. Najmudin, S.G. Preston, J.S. Wark: Laser Part. Beams **13**, 19 (1995)
220. T.E. Glover, J.K Crane, M.D. Perry, R.W. Lee, R.W. Falcone: Phys. Rev. Lett. **75**, 445 (1995)
221. T.E. Glover, J.K Crane, M.D. Perry, R.W. Lee, R.W. Falcone: Phys. Rev. E **57**, 982 (1998)
222. P.B. Corkum, N.H. Burnett, F. Brunel: Phys. Rev. Lett. **62**, 1259 (1989)
223. M.V. Ammosov, N.B. Delone, V.P. Kraïnov: Sov. Phys. JETP **64**, 1191 (1987)
224. B.M. Penetrante, J.N. Bardsley: Phys. Rev. A **43**, 3100 (1991)
225. Y. Nagata, K. Midorikawa, S. Kubodera, M. Obara, H. Tashiro, K. Toyoda: Phys. Rev. Lett. **71**, 3774 (1993)
226. T.D. Donnelly, L. Da Silva, R.W. Lee, S. Mrowka, M. Hofer, and R.W. Falcone: J. Opt. Soc. Am. B **13**, 185 (1996)
227. K.M. Krushelnick, W. Tighe, S. Suckewer: J. Opt. Soc. Am. B **13**, 306 (1996)

228. D.V. Korobkin, C.H. Nam, S. Suckewer, A. Goltsov: Phys. Rev. Lett. 77, 5206 (1996)
229. B.E. Lemoff, C.P.J. Barty, S.E. Harris: Opt. Lett. **19**, 369 (1994)
230. B.E. Lemoff, G.Y. Yin, C.L. Gordon III, C.P.J. Barty, S.E. Harris: Phys. Rev. Lett. **74**, 1574 (1995)
231. C. Rischel, A. Rousse, I. Uschmann, P-A. Albouy, J-P. Geindre, P. Audebert, J-C. Gauthier, E. Förs. Martin, A. Antonetti: Nature **390**, 490 (1997)
232. M.A. Duguay, P.M. Rentzepis: Appl. Phys. Lett. **10**, 350 (1967)
233. W.T. Silfvast, J.J. Macklin, O.R. Wood II: Opt. Lett. **8**, 551 (1983)
234. H.C. Kapteyn, R.W. Lee, R.W. Falcone: Phys. Rev. Lett. **57**, 2939 (1986)
235. H.C. Kapteyn, R.W. Falcone: Phys. Rev. A **37**, 2033 (1988)
236. D.J. Walker, C.P.J. Barty, G.Y. Yin, J.F. Young, S.E. Harris: Opt. Lett. **12**, 894 (1987)
237. M.H. Sher, J.J. Macklin, J.F. Young, S.E. Harris: Opt. Lett. **12**, 891 (1987)
238. S.A. Mani, H.A. Hyman, J.D. Daugherty: J. Appl. Phys. **47**, 3099 (1976)
239. S.E. Harris, J.F. Young: J. Opt. Soc. Am. B **4**, 547 (1987)
240. M.M. Murnane, H.C. Kapteyn, R.W. Falcone: Phys. Rev. Lett. **62**, 155 (1989)
241. M. Chaker, J.C. Kieffer, J.P. Matte, H. Pépin, P. Audebert, P. Maine, D. Strickland, P. Bado, G. Mourou: Phys. Fluids B **3**, 167 (1991)
242. J.C. Kieffer, M. Chaker, J.P. Matte, H. Pépin, C.Y. Côté, Y. Beaudoin, T.W. Johnston, C.Y. Chien, S. Coe, G. Mourou, O. Peyrusse: Phys. Fluids B **5**, 2676 (1993)
243. J.C. Kieffer, M. Chaker: J. X-Ray Science Tech. 312 (1994)
244. A. Rousse, P. Audebert, J.P. Geindre, F. Falliès, J.C. Gauthier, A. Mysyrowicz, G. Grillon, A. Antonetti: Phys. Rev. E **50**, 2200 (1994)
245. D. Umstadter, J. Workman, A. Maksimchuk, X. Liu, U. Ellenberger, J.S. Coe, C.Y. Chien: J. Quant. Spectrosc. Radiat. Transfer **54**, 401 (1995)
246. J. Workman, A. Maksimchuk, X. Liu, U. Ellenberger, J.S. Coe, C.Y. Chien, D. Umstadter: Phys. Rev. Lett. **75**, 2324 (1995)
247. J. Workman, A. Maksimchuk, X. Liu, U. Ellenberger, J.S. Coe, C.Y. Chien, D. Umstadter: J. Opt. Soc. Am. B **13**, 125 (1996)
248. J.F. Pelletier, M. Chaker, J.C. Kieffer: Opt. Lett. **21**, 1040 (1996)
249. J.F. Pelletier, M. Chaker, J.C. Kieffer: Appl. Phys. Lett. **69**, 2172 (1996)
250. J.F. Pelletier, M. Chaker, J.C. Kieffer: J. Appl. Phys. **81**, 5980 (1997)
251. G. Malka, N. Blanchot, D. Desenne, M. Louis-Jacquet, A. Mens, J.L. Miquel, O. Peyrusse: J. Opt. Soc. Am. B **14**, 2091 (1997)
252. D. Altenberger, U. Teubner, P. Gibbon, E. Förster, P. Audebert, J.P. Geindre, J.C. Gauthier, G. Grillon, A. Antonetti: J. Phys. B: At. Mol. Opt. Phys. **30**, 3969 (1997)
253. C.Y. Côté, J.C. Kieffer, Z. Jiang, A. Ikhlef, H. Pépin: J. Phys. B: At. Mol. Opt. Phys. **31**, L883 (1998)
254. J. Yu, Z. Jiang, J.C. Kieffer, A. Krol: Phys. Plasmas **6**, 1318 (1999)
255. M.M. Murnane, H.C. Kapteyn, S.P. Gordon, J. Bokor, E.N. Glytsis, R.W. Falcone: Appl. Phys. Lett. **62**, 1068 (1993)
256. S.P. Gordon, T. Donnelly, A. Sullivan, H. Hamster, R.W. Falcone: Opt. Lett. **19**, 484 (1994)
257. H.C. Kapteyn: Appl. Opt. **31**, 4931 (1992)
258. A.V. Borovsky and V.B. Mokrov: Proc. Gen. Phys. Inst. Russ. Acad. Sci. **50**, 147 (1995)
259. Y. Li, H. Schillinger, C. Ziener, R. Sauerbrey: Opt. Commun. **144**, 118 (1997)
260. A. L'Huillier, K.J. Schafer, K.C. Kulander: J. Phys. B: At. Mol. Opt. Phys. **24**, 3315 (1991)
261. Ph. Balcou, A. L'Huillier: Phys. Rev. A **47**, 1447 (1993)

262. A. L'Huillier, Ph. Balcou: Phys. Rev. Lett. **70**, 774 (1993)
263. A. L'Huillier, K.J. Schafer, K.C. Kulander: Phys. Rev. Lett. **66**, 2200 (1991)
264. J.J. Macklin, J.D. Kmetec, C.L. Gordon III: Phys. Rev. Lett. **70**, 766 (1993)
265. N. Sarukura, K. Hata, T. Adachi, R. Nodomi, M. Watanabe, and S. Watanabe: Phys. Rev. A **43**, 1669 (1991)
266. K. Kondo, T. Tamida, Y. Nabekawa, S. Watanabe: Phys. Rev. A **49**, 3881 (1994)
267. J.L. Krause, K.J. Schafter, K.C. Kulander: Phys. Rev. Lett. **68**, 3535 (1992)
268. T. Ditmire, J.K. Crane, H. Nguyen, L.B. DaSilva, M.D. Perry: Phys. Rev. A **51**, R902 (1991)
269. S. Watanabe, K. Kondo, Y. Nabekawa, A. Sagisaka, Y. Kobayashi: Phys. Rev. Lett. **73**, 2692 (1994)
270. M.D. Perry, G. Mourou: Science **264**, 917 (1994)
271. Z. Chang, A. Rundquist, H. Wang, M. Murnane, H.C. Kapteyn: Phys. Rev. Lett. **79**, 2967 (1997)
272. Ch. Spielmann, N.H. Burnett, S. Sartania, R. Koppitsch, M. Schnürer, C. Kan, M. Lenzner, P. Wobrauschek, F. Krausz: Science **278**, 661 (1997)
273. T.D. Donelly, T. Ditmire, K. Neuman, M.D. Perry, R.W. Falcone: Phys. Rev. Lett. **76**, 2472 (1996)
274. S. Kohlweyer, G.D. Tsakiris, C.G. Wahlstrom, C. Tillman, I. Mercer: Opt. Commun. **117**, 431 (1995)
275. D. von-der-Linde, T. Engers, G. Jenke, P. Agostini, G. Grillon, E. Nibbering, A. Mysyrowicz, A. Antonetti: Phys. Rev. A **52**, R25 (1995)
276. P.A. Norreys, M. Zepf, S. Moustaïzis, A.P. Fews, J. Zhang, P. Lee, M. Bakarezos, C.N. Danson, A. Dyson, P. Gibbon, P. Loukakos, D. Neely, F.N. Walsh, J.S. Wark, A.E. Dangor: Phys. Rev. Lett. **76**, 1832 (1996)
277. D.M. Chambers, P.A. Norreys, A.E. Dangor, R.S. Marjoribanks, S. Moustaïzis, D. Neely, J.S. Wark, I. Watts, M. Zepf: Opt. Commun. **148**, 289 (1998)
278. P.B. Corkum: Phys. Rev. Lett. **71**, 1994 (1993)
279. M. Lewenstein, Ph. Balcou, M. Y. Ivanov, A. L'Huillier, and P.B. Corkum: Phys. Rev. A **49**, 2117 (1994)
280. T. Ditmire, E.T. Gumbrell, R.A. Smith, J.W.G. Tisch, D.D. Meyerhofer, M.H.R. Hutchinson: Phys. Rev. Lett. **77**, 4756 (1996)
281. T. Ditmire, E.T. Gumbrell, R.A. Smith, J.W.G. Tisch, D.D. Meyerhofer, M.H.R. Hutchinson: Appl. Phys. B **65**, 313 (1997)
282. P. Salières, A. L'Huillier, M. Lewenstein: Phys. Rev. Lett. **74**, 3776 (1995)
283. M. Bellini, C. Lyngå, A. Tozzi, M. B. Gaarde, T.W. Hänsch, A. L'Huillier, C.-G. Wahlström: 1998, Phys. Rev. Lett. **81**, 297 (1998)
284. P. Salières, A. L'Huillier, Ph. Antoine, M. Lewenstein: Adv. At. Mol. Opt. Phys., **41,** 83 (1999)
285. L. Le Déroff, P. Salières, B. Carré: Opt. Lett. **23**, 1544 (1998)
286. J. M. Schins, P. Breger, P. Agostini, R. C. Constantinescu, H. G. Muller, G. Grillon, A. Antonetti, A. Mysyrowicz: Phys. Rev. Lett. **73**, 2180 (1994)
287. J. M. Schins, P. Breger, P. Agostini, R. C. Constantinescu, H. G. Muller, G. Grillon, A. Antonetti, A. Mysyrowicz: Phys. Rev. A **52**, 1272 (1995)
288. T. E. Glover, R. W. Schoenlein, A. H. Chin, C. V. Shank: Phys. Rev. Lett. **76**, 2468 (1996)
289. A. Bouhal, R. Evans, G. Grillon, A. Mysyrowicz, P. Breger, P. Agostini, R. C. Constantinescu, H. G. Muller, D. von-der-Linde: J. Opt. Soc. Am. B **14**, 950 (1997)
290. W. Theobald, R. Häbner, C. Wülker, R. Sauerbrey: Phys. Rev. Lett. **77**, 298 (1996)

291. S.G. Preston, A. Sanpera, M. Zepf, W.J. Blyth, C.G. Smith, J.S. Wark, M.H. Key, K. Burnett, M. Nakai, D. Neely, A.A. Offenberger: Phys. Rev. A **53**, R31 (1996)
292. K. Krushelnick, W. Tighe, S. Suckewer: J. Opt. Soc. Am. B **14**, 1687 (1997)
293. E. Esarey, A. Ting, P. Sprangle, D. Umstadter, X. Liu: IEEE Trans. Plasma Sci. **21**, 95 (1993)
294. X. Liu, D. Umstadter, E. Esarey, A. Ting: IEEE Trans. Plasma Sci. **21**, 90 (1993)
295. V. Malka, A. Modena, Z. Najmudin, A.E. Dangor, C.E. Clayton, K.A. Marsh, C. Joshi, C. Danson, D. Neely, F.N. Walsh: Phys. Plasmas **4**, 1127 (1997)
296. K. Krushelnick, A. Ting, H.R. Burris, A. Fisher, C. Manka, E. Esarey: Phys. Rev. Lett. **75**, 3681 (1995)
297. T. Auguste, P. Monot, G. Mainfray, C. Manus, S. Gary, M. Louis-Jacquet: Opt. Commun. **105**, 292 (1994)
298. S.-Y. Chen, A. Macksimchuk, D. Umstadter: Nature **396**, 653 (1998)
299. E. Esarey, S.K. Ride, P. Sprangle: Phys. Rev. E **48**, 3003 (1993)
300. R.L. Carman, D.W. Forslund, J.M. Kindel: Phys. Rev. Lett. **46**, 29 (1981)
301. R.L. Carman, C.K. Rhodes, R.F. Benjamin: Phys. Rev. A **24**, 2649 (1981)
302. P. Gibbon: Phys. Rev. Lett. **76**, 50 (1996)
303. L.M. Gorbunov, V.I. Kirsanov: Sov. Phys. JETP **66**, 290 (1987)
304. E. Esarey, A. Ting, P. Sprangle, G. Joyce: Comments Plasma Phys. Controlled Fusion **12**,191 (1989)
305. H. Hamster, A. Sullivan, S. Gordon, W. White, R.W. Falcone: Phys. Rev. Lett. **71**, 2725 (1993)
306. H. Hamster, A. Sullivan, S. Gordon, R.W. Falcone: Phys. Rev. E **49**, 671 (1994)
307. J.R. Marquès, J.P. Geindre, F. Amiranoff, P. Audebert, J.C. Gauthier, A. Antonetti, G. Grillon: Phys. Rev. Lett. **76**, 3566 (1996)
308. C.W. Siders, S.P. Leblanc, A. Babine, A. Stepanov, A. Sergeev, T. Tajima, M.C. Downer: IEEE Trans. Plasma Sci. **24**, 301 (1996)
309. C.W. Siders, S.P. Leblanc, D. Fisher, T. Tajima, M.C. Downer, A. Babine, A. Stepanov, A. Sergeev: Phys. Rev. Lett. **76**, 3570 (1996)
310. J.P. Geindre, P. Audebert, A. Rousse, F. Fallies, J.C. Gauthier, A. Mysyrowicz, A. Dos Santos, G. Hammoniaux, A. Antonetti: Opt. Lett. **19**, 1997 (1994)
311. J.R. Marquès, F. Dorchies, P. Audebert, J.P. Geindre, F. Amiranoff, J.C. Gauthier, G. Hammoniaux, A. Antonetti, P. Chessa, P. Mora, T.M. Antonsen, Jr.: Phys. Rev. Lett. **78**, 3463 (1997)
312. J.R. Marquès, F. Dorchies, F. Amiranoff, P. Audebert, J.C. Gauthier, J.P. Geindre, A. Antonetti, T.M. Antonsen, Jr., P. Chessa, P. Mora: Phys. Plasmas **5**, 1162 (1998)
313. J.M. Dawson: Phys. Rev. **113**, 383 (1959)
314. T. Tajima, J.M. Dawson: Phys. Rev. Lett. **43**, 267 (1979)
315. C.E. Clayton, K.A. Marsh, A. Dyson, M. Everett, A. Lal, W.P. Leemans, R. Williams, C. Joshi: Phys. Rev. Lett. **70**, 37 (1993)
316. F. Amiranoff, M. Laberge, J.R Marquès, F. Moulin, E. Fabre, B. Cros, G. Matthieussent, P. Benkeiri, F. Jacquet, J. Meyer, Ph. Miné, C. Stenz, P. Mora: Phys. Rev. Lett. **68**, 3710 (1992)
317. C.E. Clayton, C. Joshi, C. Darrow, D. Umstadter: Phys. Rev. Lett. **54**, 2343 (1985)
318. N.A. Ebrahim, P. Lavigne, S. Aithal: IEEE Trans. Nucl. Sci. **32**, 3539 (1985)

319. Y. Kitagawa, T. Matsumoto, T. Minamihata, K. Sawai, K. Matsuo, K. Mima, K. Nishihara, H. Azechi, K.A. Tanaka, H. Takabe, S. Nakai: Phys. Rev. Lett. **68**, 48 (1992)
320. M. Everett, A. Lal, D. Gordon, C.E. Clayton, K.A. Marsh, C. Joshi: Nature **368**, 527 (1994)
321. N.A. Ebrahim: J. Appl. Phys. **76**, 7645 (1994)
322. F. Amiranoff, D. Bernard, B. Cros, F. Jacquet, G. Matthieussent, Ph. Miné, P. Mora, J. Morillo, F. Moulin, A.E. Specka, C. Stenz: Phys. Rev. Lett. **74**, 5220 (1995)
323. F. Amiranoff, S. Baton, D. Bernard, B. Cros, D. Descamps, F. Dorchies, F. Jacquet, V. Malka, J.R. Marquès, G. Matthieussent, Ph. Miné, A. Modena, P. Mora, J. Morillo, Z. Najmudin: Phys. Rev. Lett. **81**, 995 (1998)
324. F. Amiranoff, D. Bernard, B. Cros, F. Jacquet, G. Matthieussent, J.R. Marques, P. Mine, P. Mora, A. Modena, J. Morillo, F. Moulin, Z. Naajmudin, A.E. Specka, C. Stenz: IEEE Trans. Plasma Sci. **24**, 296 (1996)
325. W.B. Mori, C. Joshi, J.M. Dawson, D.W. Forslund, J.M. Kindel: Phys. Rev. Lett. **72**, 1482 (1994)
326. T.M. Antonsen, Jr, P. Mora: Phys. Rev. Lett. **69**, 5220 (1995)
327. N.E. Andreev, L.M. Gorbunov, V.I. Kirsanov, A. Pogosova, R.R. Ramazashvili: JETP Lett. **55**, 571 (1992)
328. E. Esarey, J. Krall, P. Sprangle: Phys. Rev. Lett. **72**, 2887 (1994)
329. C.A Coverdale, C.B Darrow, C.D. Decker, W.B. Mori, K.C. Tzeng, K.A. Marsh, C.E. Clayton, C. Joshi: Phys. Rev. Lett. **74**, 4659 (1995)
330. K. Nakajima, D. Fisher, T. Kawakubo, H. Nakanishi, A. Ogata, Y. Kato, Y. Kitagawa, R. Kodama, K. Mima, H. Shiraga, K. Suzuki, K. Yamakawa, T. Zhang, Y. Sakawa, T. Shiji, Y. Nishida, N. Yugami, M. Downer, T. Tajima: Phys. Rev. Lett. **74**, 4428 (1995)
331. A. Modena, Z. Najmudin, A.E. Dangor, C.E. Clayton, K.A. Marsh, C. Joshi, V. Malka, C.B. Darrow, C. Danson, D. Neely, F.N. Walsh: Nature **377**, 606 (1995)
332. A. Modena, Z. Najmudin, A.E. Dangor, C.E. Clayton, K.A. Marsh, C. Joshi, V. Malka, C.B. Darrow, C. Danson: IEEE Trans. Plasma Sci. **24**, 289 (1996)
333. A.Ting, C.I. Moore, K. Krushelnick, C. Manka, E. Esarey, P. Sprangle, R. Hubbard, H.R. Burris, R. Fisher, M. Baine: Phys. Plasmas **4**, 1889 (1997)
334. C.I. Moore, A. Ting, K. Krushelnick, E. Esarey, R.F. Hubbard, B. Hafizi, H.R. Burris, C. Manka, P. Sprangle: Phys. Rev. Lett. **79**, 3909 (1997)
335. D. Gordon, K.C. Tzeng, C.E. Clayton, A.E. Dangor, V. Malka, K.A. Marsh, A. Modena, W.B. Mori, P. Muggli, Z. Najmudin, D. Neely, C. Danson, C. Joshi: Phys. Rev. Lett. **80**, 2133 (1998)
336. S.P. Leblanc, M.C. Downer, R. Wagner, S.-Y. Chen, A. Maksimchuk, G. Mourou, D. Umstadter: Phys. Rev. Lett. **77**, 5381 (1996)
337. R. Wagner, S.-Y. Chen, A. Maksimchuk, D. Umstadter: Phys. Rev. Lett. **78**, 3125 (1997)
338. L. Gorbunov, P. Mora, T.M. Antonsen, Jr.: Phys. Rev. Lett. **76**, 2495 (1996)
339. A.R. Bell, P. Gibbon: Plasma Phys. Controlled Fusion **30**, 1319 (1988)
340. G. Miano: Phys. Scr. **T30**, 198 (1990)
341. A. Pukhov, J. Meyer-ter-Vehn: Phys. Rev. Lett. **76**, 3975 (1996)
342. E.S. Weibel: Phys. Rev. Lett. **2**, 83 (1959)
343. M. Borghesi, A.J. Mackinnon, L. Barringer, R. Gaillard, L.A. Gizzi, C. Meyer, O. Willi, A. Pukhov, J. Meyer-ter-Vehn: Phys. Rev. Lett. **80**, 5137 (1998)
344. M.D.J. Burgess, B. Luther-Davies, K.A. Nugent: Phys. Fluids **28**, 2286 (1985)
345. J.A. Stamper, B.H. Ripin: Phys. Rev. Lett. **34**, 138 (1975)
346. J.A. Stamper: Laser Part. Beams **9**, 841 (1991)

347. A. Pukhov, J. Meyer-ter-Vehn: Phys. Rev. Lett. **79**, 2686 (1997)
348. M. Borghesi, A.J. MacKinnon, A.R. Bell, R. Gaillard, O. Willi: Phys. Rev. Lett. **81**, 112 (1998)
349. A.R. Bell, F.N. Beg, Z. Chang, A.E. Dangor, C.N. Danson, C.B. Edwards, A.P. Fews, M.H.R. Hutchinson, S. Luan, P. Lee, P.A. Norreys, R.A. Smith, P.F. Taday, F. Zhou: Phys. Rev. E **48**, 2087 (1993)
350. Y. Pomeau, D. Quémada: C.R. Acad. Sci. Paris **264**, 517 (1967) (in French)
351. J. Deschamps, M. Fitaire, M. Lagoutte: Phys. Rev. Lett. **25**, 1330 (1970)
352. A. Sh. Abdullaev, A.A. Frolov: Sov. Phys. JETP **54**, 493 (1981)
353. A. Sh. Abdullaev, A.A. Frolov: JETP Lett. **33**, 101 (1981)
354. T. Lehner: Phys. Scr. **49**, 704 (1994)
355. S. Eliezer, Y. Paiss, H. Strauss: Phys. Lett. A **164**, 416 (1992)
356. E. Kolka, S. Eliezer, Y. Paiss: Phys. Lett. A **180**, 132 (1993)
357. E. Kolka, S. Eliezer, Y. Paiss: Laser Part. Beams **13**, 83 (1995)
358. V. Yu Bychenkov, V.I. Demin, V.T. Tikhonchuk: Sov. Phys. JETP **78**, 62 (1994)
359. Z.M. Sheng, J. Meyer-ter-Vehn: Phys. Rev. E **54**, 1833 (1996)
360. A.D. Steiger, C.H. Wood: Phys. Rev. A **5**, 1467 (1972)
361. J.J. Thomson: *Notes on Recent Researches in Electricity and Magnetism* (Clarendon Press, Oxford, U.K. 1893)
362. A.H. Compton: Phys. Rev. **22**, 409 (1923)
363. G. Breit, J.A. Wheeler: Phys. Rev. **46**, 1087 (1934)
364. C.I. Moore, J.P. Knauer, D.D. Meyerhofer: Phys. Rev. Lett. **74**, 2439 (1995)
365. D.D. Meyerhofer, J.P. Knauer, S.J. McNaught, C.I. Moore: J. Opt. Soc. Am B **13**, 113 (1996)
366. G. Malka, E. Lefebvre, J-L Miquel: Phys. Rev. Lett. **78**, 3314 (1997)
367. C. Bula, K.T. McDonald, E.J. Prebys, C. Bamber, S. Boege, T. Kotseroglou, A.C. Melissinos, D.D. Meyerhofer, W. Ragg, D.L. Burke, R.C. Field, G. Horton-Smith, A.C. Odian, J.E. Spencer, D. Walz, S.C. Berridge, W.M. Bugg, K. Shmakov, A.W. Weidemann: Phys. Rev. Lett. **76**, 3116 (1996)
368. D.L. Burke, R.C. Field, G. Horton-Smith, J.E. Spencer, D. Walz, S.C. Berridge, W.M. Bugg, K. Shmakov, A.W. Weidemann, C. Bula, K.T. McDonald, E.J. Prebys, C. Bamber, S.J. Boege, T. Koffas, T. Kotseroglou, A.C. Melissinos, D.D. Meyerhofer, D.A. Reis, W. Ragg: Phys. Rev. Lett. **79**, 1626 (1997)
369. D.D. Meyerhofer: IEEE J. Quantum. Electron. **33**, 1935 (1997)
370. A.I. Nikishov, V.I. Ritus: Sov. Phys. JETP **19**, 529 (1964)
371. N.B. Narozhnyi, A.I. Nikishov, V.I. Ritus: Sov. Phys. JETP **20**, 622 (1965)
372. W.P. Leemans, R.W. Schoenlein, P. Volfbeyn, A.H. Chin, T.E. Glover, P. Balling, M. Zolotorev, K.J. Kim, S. Chattopadhyay, C.V. Shank: Phys. Rev. Lett. **77**, 4182 (1996)
373. W.P. Leemans, R.W. Schoenlein, P. Volfbeyn, A.H. Chin, T.E. Glover, P. Balling, M. Zolotorev, K.J. Kim, S. Chattopadhyay, C.V. Shank: IEEE J. Quantum. Electron. **33**, 1925 (1997)
374. M. Tabak, J. Hammer, M.E. Glinsky, W.L. Kruer, S.C. Wilks, J. Woodworth, E.M. Campbell, M.D. Perry: Phys. Plasmas **1**, 1626 (1994)
375. E. Lefebvre, G. Bonnaud: Phys. Rev. Lett. **74**, 2002 (1995)
376. G. Malka, J.L. Miquel: Phys. Rev. Lett. **77**, 75 (1996)
377. U. Teubner, I. Uschmann, P. Gibbon, D. Altenbernd, E. Förster, T. Feurer, W. Theobald, R. Sauerbrey, G. Hirst, M.H. Key, J. Lister, D. Neely: Phys. Rev. E **54**, 4167 (1996)

378. T. Feurer, W. Theobald, R. Sauerbrey, I. Uschmann, D. Altenbernd, U. Teubner, P. Gibbon, E. Förster, G. Malka, J-L. Miquel: Phys. Rev. E **56**, 4608 (1997)
379. G. Malka, J. Fuchs, F. Amiranoff, S.D. Baton, R. Gaillard, J.L. Miquel, H. Pépin, C. Rousseaux, G. Bonnaud, M. Busquet, L. Lours: Phys. Rev. Lett. **79**, 2053 (1997)
380. M.H. Key, M.D. Cable, T.E. Cowan, K.G. Estabrook, B.A. Hammel, S.P. Hatchett, E.A. Henry, D.E. Hinkel, J.D. Kilkenny, J.A. Koch, W.L. Kruer, A.B. Langdon, B.F. Lasinski, R.W. Lee, B.J. MacGowan, A. MacKinnon, J.D. Moody, M.J. Moran, A.A. Offenberger, D.M. Pennington, M.D. Perry, T.J. Phillips, T.C. Sangster, M.S. Singh, M.A. Stoyer, M. Tabak, G.L. Tietbohl, M. Tsukamoto, K. Wharton, S.C. Wilks: Phys. Plasmas **5**, 1966 (1998)
381. K.B. Wharton, S.P. Hatchett, S.C. Wilks, M.H. Key, J.D. Moody, V. Yanovsky, A.A. Offenberger, B.A. Hammel, M.D. Perry, C. Joshi: Phys. Rev. Lett. **81**, 822 (1998)
382. T.W. Phillips, M.D. Cable, T.E. Cowan, S.P. Hatchett, E.A. Henry, M.H. Key, M.D. Perry, T.C. Sangster, M.A. Stoyer: Rev. Sci. Instrum. **70**, 1213 (1999)
383. M. Tatarakis, J.R. Davies, P. Lee, P.A. Norreys, N.G. Kassapakis, F.N. Beg, A.R. Bell, M.G. Haines, A.E. Dangor: Phys. Rev. Lett. **81**, 999 (1998)
384. T.A. Hall, S. Ellwi, D. Batani, A. Bernardinello, V. Masella, M. Koenig, A, Benuzzi, J. Krishnan, F. Pisani, A. Djaoui, P. Norreys, D. Neely, S. Rose, M.H. Key, P. Fews: Phys. Rev. Lett. **81**, 1003 (1998)
385. A.P. Fews, P.A. Norreys, F.N. Beg, A.R. Bell, A.E. Dangor, C.N. Danson, P. Lee, S.J. Rose: Phys. Rev. Lett. **73**, 1801 (1994)
386. F.N. Beg, A.R. Bell, A.E. Dangor, C.N. Danson, A.P. Fews, M.E. Glinsky, B.A. Hammel, P. Lee, P.A. Norreys, M. Tatarakis: Phys. Plasmas **4**, 447 (1997)
387. X. Liu, D. Umstadter: Phys. Rev. Lett. **69**, 1935 (1992)
388. O. Peyrusse, M. Busquet, J.C. Kieffer, Z. Jiang, C.Y. Côté: Phys. Rev. Lett. **75**, 3862 (1995)
389. P.E. Young, P.R Bolton: Phys. Rev. Lett. **77**, 4556 (1996)
390. M. Borghesi, A.J. Mackinnon, L. Barringer, R. Gaillard, L.A. Gizzi, C. Meyer, O. Willi, A. Pukhov, J. Meyer-ter-Vehn: Phys. Rev. Lett. **78**, 879 (1997)
391. J. Fuchs, G. Malka, J.C. Adam, F. Amiranoff, S.D. Baton, N. Blanchot, A. Héron, G. Laval, J.L. Miquel, P. Mora, H. Pépin, C. Rousseaux: Phys. Rev. Lett. **80**, 1658 (1998)
392. A.J. Mackinnon, M. Borghesi, R. Gaillard, G. Malka, O. Willi, A.A. Offenberger, A. Pukhov, J. Meyer-ter-Vehn, B. Canaud, J-L. Miquel, N. Blanchot: Phys. Plasmas **6**, 2185 (1999)
393. F. Giulietti, L.A. Gizzi, A. Giulietti, A. Macchi, D. Teychenné, P. Chessa, A. Rousse, G. Cheriaux, J.P. Chambaret, G. Darpentigny: Phys. Rev. Lett. **79**, 3194 (1997)
394. J. Fuchs, J.C. Adam, F. Amiranoff, S.D. Baton, P. Gallant, L. Gremillet, A. Héron, J.C. Kieffer, G. Laval, G. Malka, J-L. Miquel, P. Mora, H. Pépin, C. Rousseaux: Phys. Rev. Lett. **80**, 2326 (1998)
395. J. Fuchs, J.C. Adam, F. Amiranoff, S.D. Baton, N. Blanchot, P. Gallant, L. Gremillet, A. Héron, J.C. Kieffer, G. Laval, G. Malka, J-L. Miquel, P. Mora, H. Pépin, C. Rousseaux: Phys. Plasmas **6**, 2569 (1999)
396. M.P. Kalashnikov, P.V. Nickles, Th. Schlegel, M. Schnürer, F. Billhardt, I. Will, W. Sandner: Phys. Rev. Lett. **73**, 260 (1994)
397. R. Kodama, K. Takahashi, K.A. Tanaka, M. Tsukamoto, H. Hashimoto, Y. Kato, K. Mima: Phys. Rev. Lett. **77**, 4906 (1996)

398. M. Zepf, M. Casto-Colin, D. Chambers, S.G. Preston, J.S. Wark, J. Zhang, C.N. Danson, D. Neely, P.A. Norrey, A.E. Dangor, A. Dyson, P. Lee, A.P. Fews, P. Gibbon, S. Moustaïzis, M.H. Key: Phys. Plasmas **3**, 3242 (1996)
399. J. Fuchs, J.C. Adam, F. Amiranoff, S.D. Baton, N. Blanchot, P. Gallant, L. Gremillet, A. Héron, J.C. Kieffer, G. Laval, G. Malka, J-L. Miquel, P. Mora, H. Pépin, C. Rousseaux: Phys. Plasmas **6**, 2563 (1999)
400. R. Kodama, K.A. Tanaka, T. Yamanaka, Y. Kato, Y. Kitagawa, H. Fujita, T. Kanabe, N. Izumi, K. Takahashi, H. Habara, K. Okada, M. Iwata, T. Matsushita, K. Mima: Plasma Phys. Controlled Fusion **41**, A419 (1999)
401. R. Trebino, D.J. Kane: J. Opt. Soc. Am. A **10**, 1101 (1993)
402. J.P. Chambaret, C. Leblanc, A. Antonetti, G. Cheriaux, P.F. Curley, G. Darpentigny, F. Salin: Opt. Lett. **21**, 1921 (1996)
403. D.Teychenné, A. Giulietti, D. Giulietti, L.A. Gizzi: Phys. Rev. E **58**, R1245 (1998)
404. S.C. Wilks, J.M. Dawson, W.B. Mori: Phys. Rev. Lett. **61**, 337 (1988)
405. P. Gibbon, E. Förtser: Plasma Phys. Controlled Fusion **38**, 769 (1996)
406. J.D. Kmetec, C.L. Gordon III, J.J Macklin, B.E. Lemoff, G.S. Brown, S.E. Harris: Phys. Rev. Lett. **68**, 1527 (1992)
407. M.D. Perry, J.A. Sefcik, T. Cowan, S. Hatchett, A. Hunt, M. Moran, D. Pennington, R. Snavely, S.C. Wilks: Rev. Sci. Instrum. **70**, 265 (1999)
408. P.A. Norreys, M. Santala, E. Clark, M. Zepf, I. Watts, F.N. Beg, K. Krushelnick, M. Tatarakis, X. Fang, P. Graham, T. McCanny, R.P. Singhal, K.W.D. Ledingham, A. Creswell, D.C.W. Sanderson, J. Magill, A. Machacek, J.S. Wark, R. Allott, B. Kennedy, D. Neely: Phys. Plasmas **6**, 2150 (1999)
409. S.J. Gitomer, R.D. Jones, F. Begay, A.W. Ehler, J.F. Kephart, R. Kristal: Phys. Fluids **29**, 2679 (1986)
410. A.P. Fews, M.J. Lamb, M. Savage: Opt. Commun. **94**, 259 (1992)
411. P.A. Norreys, A.P. Fews, F.N. Beg, A.R. Bell, A.E. Dangor, P. Lee, M.B. Nelson, H. Schmidt, M Tatarakis, M.D. Cable: Plasma Phys. Controlled Fusion **40**, 175 (1998)
412. L. Disdier, J-P Garçonnet, G. Malka, J-L. Miquel: Phys. Rev. Lett. **80**, 1454 (1999)
413. R. Loveman, J. Bendahan, T. Gozani, J. Stevenson: Nucl. Instrum. Methods Phys. Res. B: Beam Interaction Mater. At. **99**, 765 (1995)
414. T. Ditmire, J. Zweiback, V.P. Yanovsky, T.E. Cowan, G. Hays, K.B. Wharton: Nature **398**, 490 (1999)
415. Y.L. Shao, T. Ditmire, J.W.G. Tisch, E. Springate, J.P. Marangos, M.H.R. Hutchinson: Phys. Rev. Lett. **77**, 3343 (1997)
416. T. Ditmire, J.W.G. Tisch, E. Springate, M.B. Mason, N. Hay, R.A. Smith, J. Marangos, M.H.R. Hutchinson: Nature **386**, 54 (1996)
417. M. Lezius, S. Dobosz, D. Normand, M. Schmidt: Phys. Rev. Lett. **80**, 261 (1998)
418. G. Pretzler, A Saemann, A. Pukhov, D. Rudolph, T. Schätz, U. Schramm, P. Thirolf, D. Habs, K. Eidmann, G.D. Tsakiris, J. Meyer-ter-Vehn, K.J. Witte: Phys. Rev. E **58**, 1165 (1998)
419. K. Krushelnick, E.L. Clark, Z. Najmudin, M. Salvati, M.I.K. Santala, M. Tatarakis, A.E. Dangor, V. Malka, D. Neely, R. Allott, and C. Danson: Phys. Rev. Lett. **83**, 737 (1999)
A. McPherson, T.S. Luk, B. Thompson, A.B. Borisov, O.B. Shiryaev, X. Chen, K. Boyer, C.K. Rhodes: Phys. Rev. Lett. **72**, 1810 (1994)
420. J.R. Lalanne, A. Ducasse, S. Kielich: *Laser Molecular Interactions: Laser Physics and Molecular Nonlinear Optics* (Wiley, NY 1996)

421. G. Schmidt: *Physics of High Temperature Plasmas* (Academic Press, New York 1979)
422. V.G. Priymak: Dr. Sci. Thesis, Mathematical Modelling Institute of the Russian Academy of Science, Moscow, Russia (1996) (in Russian)
423. F. Verhulst: *Nonlinear Differential Equations and Dynamical Systems* (Springer-Verlag, Berlin, new York1990)
424. L.D. Landau, E.M. Lifshitz: *Quantum Mechanics: Non-Relativistic Theory* (Pergamon Press, Oxford, New York 1965)
425. A.L. Galkin: Dr. Sci. Thesis, General Physics Institute of the Russian Academy of Science, Moscow, Russia (1998) (in Russan)

Index

Springer Series on
ATOMIC, OPTICAL, AND PLASMA PHYSICS

www.ingramcontent.com/pod-product-compliance
Ingram Content Group UK Ltd.
Pitfield, Milton Keynes, MK11 3LW, UK
UKHW021828190726
13853UKWH00003B/1252

* 9 7 8 3 6 6 2 0 5 2 4 3 3 *